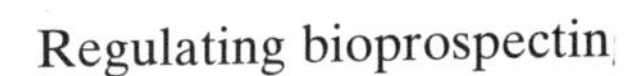

Regulating bioprospectin

Regulating bioprospecting: Institutions for drug research, access, and benefit-sharing

Padmashree Gehl Sampath

United Nations University Press
United Nations University, 53-70, Jingumae 5-chome,
Shibuya-ku, Tokyo, 150-8925, Japan
Tel: +81-3-3499-2811 Fax: +81-3-3406-7345
E-mail: sales@hq.unu.edu general enquiries: press@hq.unu.edu
http://www.unu.edu

United Nations University Office at the United Nations, New York
2 United Nations Plaza, Room DC2-2062, New York, NY 10017, USA
Tel: +1-212-963-6387 Fax: +1-212-371-9454
E-mail: unuona@ony.unu.edu

United Nations University Press is the publishing division of the United Nations University.

Cover design by Mea Rhee
Cover illustration courtesy of the National Agricultural Library, United States Department of Agriculture

Printed in the United States of America

ISBN 92-808-1112-6

Library of Congress Cataloging-in-Publication Data

Sampath, Padmashree Gehl.
Regulating bioprospecting : institutions for drug research, access, and benefit-sharing / Padmashree Gehl Sampath.
p. ; cm.
Includes bibliographical references and index.
ISBN 9280811126 (pbk.)
1. Drugs—Patents. 2. Genes—Patents. 3. Drugs—Testing—Law and legislation. 4. Genetic engineering—Law and legislation. I. Title.
K1519.B54S26 2005
344.04′233—dc22 2005011337

For Daddy and Amma

Contents

Abbreviations

BGVS	Bedrijf Geneesmiddelen Voorziening Suriname
BMS	Bristol-Myers Squibb Pharmaceutical Research Institute
CAH	Consejo Aguaruna and Huambisa
CBD (or the Convention)	Convention on Biological Diversity
CGMPs	current good manufacturing practices
CI	Conservational International
DBFs	dedicated biotechnology firms
EMEA	European Agency for the Evaluation of Medicinal Products
FDA	Food and Drug Administration (US)
GATT	General Agreement on Trade and Tariffs
GHM	Grossman-Hart-Moore
ICBGs	International Cooperative Biodiversity Groups
IDB	Inter-American Development Bank
IGC	Inter Governmental Committee (WIPO)
ILO	International Labour Organization
IPRs	intellectual property rights
ITQs	individual transferable quotas
MAT	mutually agreed terms
MFO	Mycological Facility, Oaxaca
MTA	Material Transfer Agreement
NCEs	new chemical entities
NCI	National Cancer Institute (US)
NGOs	nongovernmental organizations

NIH	National Institute of Health
NSF	National Science Foundation
OTA	Office of Technological Assessment (US)
PhRMA	Pharmaceutical Research and Manufacturers of America Foundation
PIC	Prior Informed Consent
R&D	research and development
SMEs	small and medium-sized enterprises
TCE	Transaction Cost Economics
TK	traditional knowledge
TRIPS Agreement	Trade Related Aspects of Intellectual Property Rights Agreement (1995)
UNCTAD	United Nations Conference on Trade and Development
USAID	US Agency for International Development
USDA	US Department of Agriculture
UZACHI	coalition of communities of Zapotec and Chinantec ethnicity in Mexico
WIPO	World Intellectual Property Organization
WTO	World Trade Organization

Acknowledgements

This book is the result of research conducted between 1998 and 2004, in several stages. The underlying framework is based on my doctoral research, while research at UNU-INTECH helped enrich the analysis further. I am grateful to professors Peter Behrens, Hans-Bernd Schaefer and Manfred Holler, Graduate College for Law and Economics, University of Hamburg; professors Robert Cooter, Suzanne Scotchmer, Gordon Rausser, Brian Wright, Howard Shelanski, and Daniel Rubinfeld, at the University of Berkeley at California; Professor Peter Drahos, Australian National University and Professor Joseph Straus, Max Planck Institute for Intellectual Property and Competition Law, Munich, for helpful comments and discussions during my doctoral research.

My sincere thanks to the Deutsche Forschungsgemeinschaft (DFG) for the scholarship and travel-related financial support during my doctoral research, without which collaborations with the Max Planck Institute for Intellectual Property and Competition Law, Munich, and the Center for Law and Economics, University of California and Berkeley, would not have been possible.

I am grateful to Professor Carlos M. Correa, University of Buenos Aires, Dr. Gordon Cragg, and Dr. David Newman of the Natural Products Branch, National Cancer Institute, USA, for their comments and suggestions on drafts of several chapters in this book. I thank Sarah Laird, Professor Brent O. Berlin and Graham Dutfield for their time and cooperation.

I thank my husband Dirk for his support and comments on several drafts of this book. I am grateful to my family for their unstinting encouragement during these years. I thank Sophia Gross and Eveline in de Braek, UNU-INTECH, for editorial assistance. All responsibility for errors of fact and interpretation remains with the author.

1

Bioprospecting for drug research: An overview

Introduction

Biodiversity prospecting or *bioprospecting*, which refers to the process of looking for potentially valuable genetic resources and biochemical compounds in nature (Eisner, 1991; Ried et al., 1993), is not a new phenomenon. Both natural products and traditional knowledge have contributed extensively to drug research for several centuries now. Aspirin® (derived from Willow Bark used as a painkiller), Reserpentine (from the Indian Snake Root for hypertension), D-tubercurarine (from arrow poisons used as a muscle relaxant in surgery), Artemisin (derived from *Artemisia Annua* or the Quinhaosu used as an anti-malarial agent), and Vincistrine and Vinblastine (derived from Rosy Periwinkle used as anti-cancer drugs) are some frequently cited examples of drugs discovered from natural products.

But in recent times, the increased use of natural products and traditional medicine in drugs, and the sizeable market revenues associated with such use, has renewed focus on their role in pharmaceutical drugs and botanical medicines. Enhanced market prospects have been accompanied by concerns on how the benefits are to be shared from such commercialization activities, the potential impact of the growing use of natural products in the drug industry on sustainable use and conservation of genetic resources, and the implications of this industrial activity on capacity building in developing and least developed countries.

Against these concerns, the recognition of the rights of national governments to regulate access to genetic resources (Article 15) and the rights of indigenous and local communities on their traditional knowledge, innovation, and practices (Article 8(j)) under the Convention on Biological Diversity (1993) was championed as the starting point of a fruitful discourse on the roles and responsibilities of users and providers in bioprospecting for drug research and development (hereafter, R&D).

But just over a decade after the Convention on Biological Diversity (hereafter, the CBD or the Convention), the interest in bioprospecting as a source of new drugs seems to be waning. Developing countries, which were so optimistic of making a headway into the biotechnology era through their genetic resources, are caught up with problems of enacting enforceable national regimes on access to genetic resources and traditional knowledge. With international negotiations continuing on several issues that affect bioprospecting – such as an international regime to govern access to genetic resources, a certification system to prove source of origin of genetic resources, and traditional knowledge – the legal situation too is in a constant state of flux. Drug companies, stuck between the choice of exploring newer, more promising technologies and using natural products for drug R&D amidst legal uncertainty and potentially unrealistic benefit-sharing expectations, are seemingly choosing the former option. According to recent reports, major companies such as Monsanto and Bristol-Myers Squibb have shut down their natural products divisions entirely, while Merck has discontinued collaboration with INBio of Costa Rica after a last grant of US$130,000 in 2001 (Dalton, 2004b: 599). Many other large drug companies have scaled back on bioprospecting and depend on smaller biotechnology companies for natural products-related services (Dalton, 2004a: 576).

Regardless of international mandates within the CBD and other international agreements such as the Agreement on Trade Related Aspects of Intellectual Property Rights, national laws on bioprospecting that define property rights on genetic resources and traditional knowledge and prescribe rules of contracting are the most critical factors to enable bioprospecting in a balanced way. This book is an investigation into optimal property rights structures and institutional mechanisms that can facilitate the process of bioprospecting for drug research while balancing the goals of optimal drug R&D with the diverse demands placed by recognition of rights over traditional knowledge and access to genetic resources, benefit-sharing, and biodiversity conservation. Using an interdisciplinary law and economics methodology, the focus of the analysis is on the economics of contracts in the drug R&D process using genetic resources, to show that the rights that are exchanged at each stage of the process are complementary to one another. Therfeore, attempts to define and enforce the rights on access to genetic resources and traditional knowledge in isolation with the drug R&D process result in a failure to realize the economic potential of bioprospecting for sustainable development and biodiversity conservation in source countries. These analytical results are substantiated by examples of bioprospecting collaborations in several countries and a critique of the institutional and contractual factors that led to their success or failure.

International law and the politics of bioprospecting

The coming into force of the CBD in 1993 was largely welcomed for its paradigm shift from free and unilateral exchange (from the genetic-resources-rich South to the industrialized North), to a more restricted and increasingly codified exchange of genetic resources and traditional knowledge, where exchange was made conditional on several factors, the most important one being the sharing of benefits between users and providers (Etkin, 2003).

The text of the CBD, a global framework to regulate issues of sustainable use and conservation of biological diversity, is novel in the sense that it seeks to address the problem from both ecological and economic perspectives. Its main objective is the conservation of biological diversity, the sustainable use of its components, and the fair and equitable sharing of the benefits arising out of the commercialization of genetic resources (Article 1). Identification of rights on access to genetic resources, traditional knowledge and provision for benefit-sharing are to play a role within this broader scheme. Yet, these provisions impose "fragile obligations" on drug companies to ensure that exchange of genetic resources and associated traditional knowledge occurs in a context where providers are guaranteed equitable returns (Hayden, 2003: 1). But the CBD's biggest shortcoming is that it "lacks contextualization to local ecopolitical circumstances and its integration across local, regional, commercial and environmental frontiers has been lax" (Etkin, 2003).

The vagueness of mandates contained in the CBD have posed major hurdles in reaching consensus on optimal national legal frameworks for bioprospecting. Not only have policy dialogues on the meaning and import of these provisions been polarized amongst countries of the South and the North, but they have also been rendered more complicated by the Agreement on Trade Related Aspects of Intellectual Property Rights (1995) (hereafter, the TRIPS Agreement), an annex to the Agreement that created the World Trade Organization. The TRIPS Agreement, which deals with aspects of intellectual property protection, has brought to the fore issues pertaining to the interaction between intellectual property rights on results of drug R&D and access, traditional knowledge, and benefit-sharing provisions of the CBD. As a result, regulation of bioprospecting has proceeded with an ignorance of the complexities of biotechnology-based drug research, the challenges in leveraging both genetic resources and traditional knowledge, and the need for a comprehensive approach to promote local capacity in developing countries in this regard (see for example, Miller, 1997; UNCTAD, 2000, 2001).

Access and traditional knowledge: Different stakeholder interests

The ongoing debate over what form the mandates on access and traditional knowledge ought to assume at the national levels, and how these should be reconciled with the provisions of the TRIPS Agreement, reveals different levels of interests involved in bioprospecting.[1] At the supra-national level, there is the overall global long-term interest in conserving genetic resources and related traditional knowledge as laid out by the CBD. At the national level, there are the interests of the source countries in regulating access and ensuring that the benefits accruing from commercialization of traditional knowledge are shared with its indigenous and local communities and those of user countries in ensuring efficacious access to genetic resources. Source countries' interests are further split between conservation of biological diversity and ensuring that benefits for the use of genetic resources and traditional knowledge are shared with the holders of these resources. The interests of indigenous and local communities, to be consulted in the rule-making process for access and benefit-sharing in order to ensure that the use, exchange, and benefit-sharing aspects respect their customary laws and institutions, act at the local level. But an interest permeating through all these levels that cannot be neglected is that of the scientific research and international trade which depends on proper and reliable access to genetic resources and traditional knowledge-based information. The design of successful legal instruments for the protection and realization of the economic potential of genetic resources and traditional knowledge depends on the way the first three levels of interests are balanced with the needs of drug R&D.

Stakeholder lineage has conditioned different perspectives on whether countries can reconcile their obligations under the CBD and the TRIPS Agreement to cater for issues of conservation, sustainability, and equity simultaneously.[2] Each one of these stakeholder perspectives presents simplistic interpretations of complex policy issues (Svarstad, 2000). Although several other categories of stakeholders can be delineated (see for example, Hayden, 2003), at a broad level three main perspectives operate. There are those who believe that countries can reconcile their obligations between the TRIPS Agreement and the CBD to their advantage. There are others who feel that such contracts will serve the short-term profit goals of firms more and neglect impending issues of conservation of biodiversity and equitable benefit-sharing with indigenous and local communities. Some others feel that the difficulty in negotiating such contracts is the main reason why the goals of conservation and equity will not be attained. These different perspectives depend on the stakeholder interests that parties represent and their ability (or inability) to balance their interests with the restrictions placed by the drug R&D process itself.

The "biopiracy" perspective

Shiva defines biopiracy as a process by which the rights of indigenous cultures to their genetic resources and associated traditional knowledge are replaced by monopoly rights of those who exploit these resources (Shiva, 1997: 31). The biopiracy view finds its basis in arguments of neo-imperialism, where bioprospecting is depicted as one of those processes that supports the usurpation of resources and knowledge of indigenous people by firms and other Western counterparts. "Bad patents" issued on uses of traditional knowledge such as Turmeric, Neem, and Ayahuasca (see Chapter 3) have been used by proponents of this perspective to make the case that intellectual property rights as propagated by the TRIPS Agreement contain a built-in bias against traditional knowledge and rights of indigenous communities. Stringent regimes for access and traditional knowledge have been seen as a major way of preventing such unfair enrichment.

The bioprospecting perspective

The "bioprospecting" perspective expresses optimism that through bioprospecting, all three objectives of the CBD – sustainable use, conservation of biological resources, and benefit-sharing – can be met. In this perspective, bioprospecting is seen a venue of revenue generation from potentially valuable traditional knowledge and genetic resources situated in the South. In the presence of well-designed laws and contracts, bioprospecting presents a "win-win" situation where benefits generated can be used for a range of purposes – improvement to livelihoods of indigenous and local communities, biodiversity conservation programmes and biotechnological capacity building (see, among others, Ried et al., 1993; Balick et al., 1996; Svarstad, 2000).

The sceptics

The category "sceptics" is self-explanatory to a large extent. The perspective is marked by large-scale scepticism of the impact of drug

R&D on creating incentives for biodiversity conservation in source countries (see, for example, Simpson, Sedjo, and Ried, 1996; Simpson and Sedjo, 2004). There is also a great deal of scepticism that the CBD and its unrealistic expectations on benefit-sharing may render drug R&D unprofitable.

Achieving clarity: The main issues

National bioprospecting frameworks have to create a balance between the promotion of R&D into new drugs or related products for consumers the world over, recognition of rights of traditional knowledge of indigenous and local communities, conservation of local biodiversity, and possibly harness other potential positive economic effects of such contracts, such as local capacity building.

But split between the various stakeholder perspectives, there is still a pervasive lack of clarity on fundamental issues of relevance to bioprospecting frameworks. The main issues that still need to be resolved are as follows:

Subject Matter: The nature and scope of countries' rights to regulate access to genetic resources, and how they should be reconciled with the rights of private property owners and the rights of holders of traditional knowledge are not at all clear. Assuming that the *sui generis* right on traditional knowledge should be an intellectual property right, there is confusion on what the subject matter of the right should be. There needs to be a systematic attempt to clarify the information categories that could be called "traditional knowledge" which list the plausible contribution of respective traditional knowledge categories to collaborative research precisely. Rights structures for traditional knowledge should be decided upon based on these categories.

Identification of Beneficiaries: Associated with the above-mentioned issue is the issue of identification of beneficiaries. How should benefits that accrue from access to genetic resources be split between the national government, private holders of genetic resources, and holders of traditional knowledge? Should the right to traditional knowledge be defined as broadly as possible in order that as many communities as possible are included, or should it be defined in accordance with the needs of the process of which it is a part (e.g. R&D).

The Issue of Biodiversity Conservation: It is widely acknowledged that a key reason for the recognition of countries' rights to regulate access to genetic resources is biodiversity conservation. Yet what the main functions of the right to access are, and how these can be performed to complement national biodiversity strategies, is still relatively unexplored. The same can be said of the link between revenues generated through the regulation of access to genetic resources and biodiversity conservation. Article 8(j) is also titled *in-situ* conservation. The nature of the contribution that the grant of rights on traditional knowledge could have on *in-situ* conservation efforts of communities, and the institutional mechanisms that could foster them, are also not clear.

Specific Attributes and Implementation Mechanisms: The attributes of the rights to access and traditional knowledge vary from country to country. The different features reveal a lack of clarity (or disagreement) on critical features such as the duration of the rights, implementation mechanisms and nature of overlap between these two rights themselves, and between the rights to access and traditional knowledge and other rights in the drug R&D process.

The need for a process-oriented approach to policy making

Ultimately, how rights to traditional knowledge and access should be defined depends less on the text of the CBD or the extent of flexibility allowed in the TRIPS–CBD interactions, and more on the structure of the drug industry, the economic exchange between the various actors in the R&D process, and how best these rights cater to the contractual needs of the actors.

In contrast to the simplistic view that considers bioprospecting to be a one-shot contract between the end-developer drug firm, the national access authority, and the community that possesses traditional knowledge, in reality there seem to be very few cases where large firms approach source countries directly for access. Genetic resources and traditional knowledge, wherever applicable, are sourced through a variety of channels – private individuals, specialized agencies, firms, or even botanical gardens – all of whom render such services for varying rates (Laird, 1993). Making huge upfront payments is also not the general rule due to the rampant uncertainty inherent in drug research.

Large-scale neglect of the economic exchange processes within which the rights over access and traditional knowledge have to operate have led to extremely bureaucratic and unrealistic laws that mostly view bioprospecting contracts as single strategic agreements. Such laws, which reflect neither the complexity of the contracts nor of the drug R&D process, act as a major disincentive for interested parties to explore the potential of bioprospecting.

In reality, the contract that each actor in the R&D process would like to sign is a reaction to various market imperfections and incentive problems in the market for bioprospecting. National laws and institutions for bioprospecting are key mechanisms for attaining the right balance between economic efficiency, the TRIPS Agreement, and the goals of the CBD vis-à-vis the terms and conditions for the exchange of genetic resources and traditional knowledge.

In the absence of process-oriented information on how the drug industry makes use of genetic resources and traditional knowledge, and the economics of the use process and how it may affect incentives of parties to use, trade, or conduct R&D on genetic resources, it is not possible to decide upon which property rights allocations suit the needs of the resources and the needs of the contracting parties best.

Therefore, any analysis of optimal property rights allocation for bioprospecting has to start with the drug R&D process. Incorporating the economic use processes into the analytical framework not only helps predict the optimal legal framework to regulate access and traditional knowledge, it also helps test the viability of one form of property rights structure over another. Using this, one can also answer the larger question that the multitude of national approaches to regulate bioprospecting raises: can there be many effective ways of regulating one and the same activity – namely, bioprospecting. Either these institutional mechanisms are all similar in matter and content, or they only look similar but, content-wise, there are certain key differences that affect their efficiency properties, or, finally, they are different but these differences do not matter.

Book structure and organization of chapters

To be able to predict property rights structures that can outdo others in regulating bioprospecting, the analysis begins with the stages of the drug R&D process, the parties involved, and the rights that are exchanged at every stage. Chapter 2 looks at the structure of the industry, the various actors involved, and the precise role of traditional knowledge and genetic resources in drug R&D. The exercise is to explain, from a user perspective, what these resources are worth to the drug R&D process and the contractual aspects relevant for their exchange.

Taking this as the vantage point, Chapter 3 is an analysis of the main legal provisions of the CBD and TRIPS Agreement that affect bioprospecting. The legal and political reasons that divide countries on questions of interpretation of the key provisions in both agreements are highlighted. This chapter tries to separate the main legal policy conflicts from the political controversies on bioprospecting. By doing so, the chapter shows that when the rights on access to genetic resources and traditional knowledge are considered in conjunction with the limitations that the CBD itself places on their nature and context, the main legal controversies regarding their scope and context can be resolved to a large extent. The chapter also derives key legal conclusions for operationalizing the rights to access and traditional knowledge at national levels.

Optimal legal frameworks for bioprospecting that balance the needs of the R&D process and all stakeholder interests in order to foster sustainable collaborations between communities, access authorities, and firms will require taking market realities into account in a comprehensive way. Chapter 4 is an investigation into the market imperfections in drug R&D based on genetic resources. Owing to the fact that the drug R&D process is a long and risky one involving costs and investments at each stage, these contracts are essentially incomplete.[3] The imperfect market conditions between the various actors that have to exchange these property rights create a series of transaction costs that stall or hinder the bargaining procedures. The role of law in such a market is to provide a set of default rules that facilitate optimal contractual arrangements to deal with risk-sharing and distribution of benefits, as well as the right incentives for mutual collaboration and biodiversity conservation by clearly laying down the bargaining thresholds. This would mean not only a framework of well-defined and enforceable property rights, but also appropriate institutional arrangements that provide for the optimal conditions for the exchange of these property rights through contracts. Chapter 4 analyses these market imperfections using transaction cost economics and the economic theories of contracting. Whereas several of the market imperfections enumerated in this chapter can be solved through well-defined property rights on traditional knowledge and access, many others call for institutional mechanisms that facilitate contracts.

Chapter 5 is a law and economics investigation into a well-defined and enforceable property right for traditional medicinal knowledge. Chapter 6 similarly deals with finding a precise definition of the right to access.

To show how the market imperfections related to asymmetric information and uncertainty can be solved through appropriate contractual structures, the latter half of Chapter 6 explores how access, through contractual facilitation, can help reduce these information asymmetry and uncertainty-related problems in bioprospecting contracts. Chapter 7 contains a summary of results and policy recommendations.

General usages in this work

Efficiency is used in the sense of maximizing the sum of producer and consumer surpluses minus environmental externalities of unsustainable resource usage. In this case, the producers are institutions involved in the R&D and drug production process, namely, pharmaceutical firms, intermediaries, research institutions, communities as holders of traditional knowledge, and the owners of natural resources (such as private holders, communities, or states). The consumer surplus is to be measured by the utility of the consumers the world over. The greater the number of medicinal products that can be made available at competitive prices and at better quality, the higher the utility of the consumers. Thus defined, efficiency has to take care of the wealth of every group involved, be it firms, countries, or local and indigenous communities minus the externalities generated by the entire process. If the legal rules enable contracting to take place under competitive conditions, then parties will get shares as part of the total surplus according to their respective contributions. Any inefficiency in the organization of economic activity represents an unrealized opportunity. In this definition of efficiency, distribution of surplus from economic activities plays a key role. Benefits have to be shared amongst each party that participates in the contracts, be it communities, source countries, firms, or other individuals, in accordance with their extent of contribution to the creation of contractual surplus. The contribution of any individual or group to the creation of surplus is measured by the productivity of their assets and/or efforts. Economic criteria of efficiency is also restricted by the rights that are granted. When rights are granted over certain assets, such as rights to traditional knowledge, access, or other forms of IPRs, efficiency demands that these rights are not exchanged without the permission of those in whom the rights are vested.

Biological resources and *genetic resources* are used as defined under Article 2 of the CBD.[4] In the analysis, wherever questions of conservation and sustainable use are discussed, it is done so in the broader sense of "biological resources". Access rights and traditional knowledge rights, as well as benefit-sharing issues, are discussed only for genetic resources.

Drug discovery/drug research and development is used interchangeably to denote the process of discovering, developing, and commercializing both pharmaceutical drugs and botanical medicines based on genetic resources (ten Kate and Laird, 1999).

Biotechnology denotes all technological applications that use biological systems, living organisms, or derivatives of living organisms used to make or modify products or processes for specific use (Article 2, CBD).

Source countries denote tropical countries that host genetic diversity and traditional knowledge.

User countries denote the industrialized countries that have the biotechnological capacities and have firms that are interested in accessing genetic resources and traditional knowledge for research purposes. Similarly, *User firm* implies any firm that is interested in using genetic resources or traditional knowledge for its R&D.

Notes

1. These categories are based on those identified by Biber-Klemm (1999: 7).
2. Svarstad presents two main discourses – bioprospecting and biopiracy (2000: 19). She notes that, "Each of these discourses is constituted by a message, frequent uses of certain metaphors and ways of telling stories of specific incidents of bioprospecting. A common feature, however, is that both proponents and opponents advocate values concerning the well-being of people and the environment in developing countries."
3. In economic theory, contracts are "incomplete" when all the contingencies cannot be provided for in advance. Due to the transaction costs involved, "the parties will quite rationally leave out many contingencies, taking the point of view that it is better 'to wait and see what happens' rather than cover a large number of unlikely eventualities" (Hart, 1995: 21–24).
4. Article 2 of the CBD on Use of Terms defines "Biological Resources" to include genetic resources, organisms or parts thereof, populations, or any biotic component of ecosystems with actual or potential use or value for humanity. "Genetic Resources" means genetic material of actual or potential value.

2

Drug R&D and the structure of the industry

Introduction[1]

Genetic resources – plants, animals and micro-organisms – are a valuable source of drugs either directly or indirectly by providing the lead information on which drugs are based. Biotechnological techniques have not only enhanced the precision with which natural molecular basis of diseases and disease causation can be identified but have also made it easier for drug research to identify important therapeutic agents using natural products as the basis (Rodney and Garibaldi, 2002). This has given a new impetus to considering natural products as the source of potential new drugs, despite the development of alternate drug research techniques, such as the "rational" approaches of molecular genetics, or synthetic drug design (Fellows and Scofield, 1995). This is also corroborated by the fact that natural products still continue to source a large number of new molecules useful for drug development. Newman et al. (2003: 1033) note in this context: "despite many years of efforts on the part of the pharmaceutical industry in high-throughput screening of (predominantly) combinatorial chemistry products, in the year 2000, 2001 and 2002 (which should have provided a sufficient time span for early efforts in the late 1980s and early 1990s to have produced approved new chemical entities (hereafter, NCEs)), the natural products field is still producing ~50% of all small molecules".

This chapter conducts an analysis of the structure of drug R&D processes for both pharmaceutical drugs and botanical medicines in order to understand the potential role of genetic resources and traditional knowledge, the economic exchange process, property rights that are relevant in practice and the main actors. For a book of this kind, the distinction between pharmaceutical drugs and botanical medicines is important for two reasons. Firstly, the process of drug R&D is different in the two sectors and, secondly, the role of genetic resources and traditional knowledge is also quite distinct. Hence, distinguishing between them allows us to look into the subtler issues of R&D, contractual problems, genetic resource use levels, and benefit-sharing.

Genetic resources and the drug industry

Pharmaceutical drugs of natural origin can be classified into different categories, depending on the R&D process and the extent to which the final drug is derived from natural products. Pharmaceutical drugs can be completely based on the natural product, they can be derived from the natural product (as semi-synthetic modifications), they can be biologicals or biopharmaceuticals (peptide or protein isolated from original organism/cells or produced using recombinant DNA techniques), or can be completely synthetic but modelled on a natural product inhibitor or a molecular target (Newman et al., 2003: 1024). In contrast, botanical medicines are almost always derived from the natural product on which they are based. As ten Kate and Laird note (1999: 78), botanical medicines "are produced from whole plant materials" and are a collective function of the active ingredients in the plant itself. There can be four different types of botanical medicines – raw herb material, extracts, standardized extracts, or phytomedicines (Laird and ten Kate, 2002: 259).

The market for pharmaceutical drugs

The top five pharmaceutical companies for 2003 were Pfizer (global market share 11%), GlaxoSmithKline (market share 6.9%), Merck and Co. (market share 5%), Astrazeneca (market share 4.8%) and Johnson and Johnson (market share 4.7%) (IMS, 2004a). According to retail pharmacy sales to August 2004, the top five companies were Pfizer, Glaxo SmithKline, Merck, Astrazeneca, and Novartis (IMS, 2004b). Pfizer, currently the world's largest pharmaceutical company, expects a revenue of US$53 billion at the end of 2004 (Wall Street Journal, 2004). Pharmacy sales between September 2003 and August 2004 are estimated to have grown at 8 per cent in the top thirteen markets worldwide, to a figure of US$337.7 billion (IMS, 2004a). Of the several drugs of natural origin with very high global markets (see Laird and ten Kate, 1999), Merck's Zocor® is still the drug with the second largest sales worldwide (ibid.).

Despite availability of other methods of drug discovery, natural products are constantly in demand as a source of new drugs, and this is especially the case in several disease categories such as cancer and infectious diseases (Cragg et al., 1997). As a result, a large percentage of pharmaceutical R&D and resulting drugs consist of natural product components, so much so that of the total pharmaceutical sales of US$300 billion in 1997, US$75 billion can be attributed to sales of drugs that have a natural origin (ten Kate and Laird, 1999). A survey conducted by Cragg et al. (1997) that covered all new drugs approved by the Food and Drug Administration of the USA (hereafter, FDA), and similar organizations in other countries, found that of the 87 cancer drugs approved worldwide up to the end of 1995, 67 per cent had a natural origin. Newman and Laird (1999) conducted another study that tried to pin down the impact of natural products on blockbuster drugs. Disagreeing with the figures in the earlier studies, they found that out of the 25 best-selling drugs worldwide, biologicals, natural products, or drugs derived from natural products accounted for 42 per cent of the sales, with a total value of US$17.5 billion in 1997 (ibid.: 334). Going by product types, they found that whereas the best-selling drugs were anti-ulcer medications, other categories such as cholesterol-lowering, anti-bacterials, and immunosuppressants were also amongst the top 25 selling drugs worldwide. Merck, for example, credited 50.6 per cent of its total sales in 1997 to natural product drugs or natural product-derived drugs (Laird and ten Kate, 2002: 249). More recently, Newman, Cragg and Snader (2003) expanded their earlier paper (from 1997) on the sources of drugs worldwide between 1981 and 2002. According to them, of the 1,031 NCEs that were approved between 1981 and 2002, and spread over all countries/diseases/sources, only 33 to 43 per cent of them were synthetic in nature.[2]

The market for botanical drugs

The market for botanical medicines includes those that are generally health-enhancing (also known as "nutraceuticals") as well as natural remedies for problems that potentially range from the simple to very serious illnesses – i.e. from colds, coughs, depression, fatigue, nervousness, to sleeplessness and insomnia, and even to diabetes and different kinds of cancer. Botanical medicines can either be raw herbs, tinctures, extracts (containing greater concentration of the original material), or standardized extracts (ten Kate and Laird, 1999: 80). Botanical medicines containing total extracts are products that are made out of all the extractable compounds contained in a plant, plant part, or combination of plants (ibid). In the case of botanical medicines that contain total extracts, the extracts can also be standardized. In such products, the extracts are standardized against a set of chemical markers (or biological activities) in the original material and, therefore, sometimes do not contain other compounds that were present in the original botanical material (ibid). These kinds of botanical medicines are called "phytomedicines".

The growing popularity of botanical medicines is to a large extent a consumer-driven trend and reveals a gradual shift in the "paradigm of biotherapeutics" (McCaleb, 1994; Etkin and John, 1998). The fact that many diseases that are prevalent amongst Western societies today are forms of heart ailments, types of cancer, and nervous disorders – diseases which can be devastating in impact, but the exact reason for their occurrence is unknown – has led to an emphasis on healthier living (Etkin and John, 1998: 12). In addition, since phytomedicines require scientific validation of the properties of the original genetic material for the identification and standardization of active ingredients, they find easy acceptance in developed countries (ten Kate and Laird, 1999). Botanical medicines, when based on traditional medicines, also use as a selling point such arguments of efficacy in traditional medicine as "evidence with regard to safety and experience is based on thousands of years of experience" (Ballance et al., 1992: 16).

The largest global markets for botanical drugs are in developed countries such as Germany, China, Japan, France, Italy, UK, USA, and Spain (NBI, 1998, cited in ten Kate and Laird, 1999). Market projections estimate sales in the botanical industry to be around US$20 billion globally, but the natural component is only estimated to be around US$3 million (Laird and ten Kate, 2002).

Drug research and industry structure

Pharmaceutical drug R&D is a research-intensive, high-risk, high-investment activity. The pharmaceutical industry, as we see it today, gained ground after World War II and by the end of the 1950s consolidated itself into a business that was intensive on research and advertising capabilities to promote its products (Gereffi, 1983). Today, the industry itself is likely to be worth US$200 billion or more (Walsh, 2003). Most of the world market for pharmaceutical drugs is dominated either by the life sciences or by pharmaceutical companies based in the USA, Switzerland, and the European Union (EU) – specifically, Germany and the United Kingdom – with developing countries such as India and China which are proficient in drug research being rather the exception.

Risk, investment, and uncertainty in pharmaceutical R&D

Several estimates of pharmaceutical R&D investments, total timeline required for a marketable product, risk, and fall-out rates within the drug R&D process exist. Whereas the OTA (1991) estimated the cost of "bringing a new medicine to the market" at US$231 million, a recent study by the Tufts Center (2001) that was sponsored by the Pharmaceutical Research and Manufacturers of America Foundation (PhRMA) showed that the average cost of developing a pharmaceutical drug in the USA has risen from US$54 million in 1979 (calculated at 1976 rates) to US$231 million in 1991 (calculated at 1989 rates) and to US$802 million in 2001.[3] PhRMA similarly estimated that US$500 million is spent on developing each new chemical entity, taking into account the cost of failed compounds and interest costs over the entire period of the investment (PhRMA, 2001). In practice, the real cost of developing a pharmaceutical drug may be much lower than such an estimate. This is because such estimates include not only the costs of R&D into potential leads and candidates that fail to become commercial products, but also contain the costs of borrowing money to finance the R&D process and the marketing costs of the product (which can add up to approximately 40 per cent of all total costs). Whereas the former is a part of R&D expenses, the latter is not.[4]

On the question of fall-out rates, the Office of Technological Assessment of the USA (hereafter, OTA) estimated that only one in 10,000 compounds screened may end up becoming a successful drug, only one in 10 compounds that enter clinical trials makes it to the market, and only one in every 30 per cent of marketed drugs recovers its R&D costs (OTA, 1991: 82–83).[5] Other estimates suggest that roughly one in 10,000 compounds screened ends up as a lead (Ried et al., 1993). In this context, the experience of the National Cancer Institute of the USA (hereafter, NCI) makes for interesting evidence. Between 1960 and 1982, some 35,000 plant samples (representing about 12,000 to 13,000 species) were processed at the NCI to yield 114,000 extracts. Though a significant number of interesting active chemotypes were discovered from these extracts, only two compounds advanced to the stage of development into commercial products. These were Taxol® (e.g. paclitaxel and its analogue docetaxel) and camptothecin. Although the latter proved to be too toxic in clinical trials to become a commercial drug, it has yielded commercial analogues, such as topotecan (Hycamptine®) and irinotecan (Camptosar®). One other product, homoharringtonine, has advanced through clinical trials for the treatment of refractory leukaemias. In total, thus far, 114,000 extracts derived from approximately 12,000 species have given only two compounds currently yielding products of commercial value (further derivatives and analogues of taxol and camptothecin are being developed which will probably become commercial products), with homoharrintonine being another likely product.

What follows from such information from the R&D process is that it is probably more realistic to estimate that of over 100,000 compounds screened only one compound may warrant clinical trials. In any given scenario, the chances of discovering a useful commercial drug are enhanced if initial natural product extracts are exposed to as many screens as possible.

Similarly, on the question of time-frames, it is generally estimated that it may take between 7 and 18 years from the start of any R&D programme to arrive at a marketable product (ten Kate and Laird, 1999). At the same time, there are examples such as Taxol, where the time lapse between the initial collection and the marketable product was 30 years (1962 to 1992). In the case of Lovastatin, a cholesterol lowering drug, there was a gap of about 30 years between the discovery of HMG-CoA and the approval of the drug. The lapse of time between the start of pharmaceutical R&D and the arrival of the final marketable product together with the uncertainty related to product demand mandates that the industry focuses on products for which they can predict a high future demand. Lastly, although it takes years of effort and large investments to develop, test, and market a drug, it is, in many cases, a relatively easy task to copy such drugs. This is, to a large extent, the reason why the pharmaceutical industry is so dependent on intellectual property protection, especially patents, to protect investments (Arora, 1995; Jaffe, 1999).

The changing structure of the pharmaceutical industry

The high risk and high investment nature of the drug industry, coupled with its intellectual property orientation, generally act as strong barriers to entry for newcomers. Although there are over 10,000 pharmaceutical companies worldwide, probably only around 100 are of true international significance (Walsh, 2003: 3). The industry has seen a gradual metamorphosis into oligopolistic structures with increased consolidation in recent times in order to strengthen in-house scientific capabilities for superior innovation, as well as to expand their technologies, product portfolios, and market performance. A peculiar characteristic of the pharmaceutical industry is the existence of an industry "core" composed of a relatively small group of firms that persistently hold a predominant position in the market (Bottazzi and Secchi, 2004). These include the top five identified earlier on in this chapter, followed by others such as Roche, Bayer, Merck, Schering (Schering Plough in the USA and Schering AG in the EU), Glaxosmithkline, and BMS. These presently account for a third of all global sales (Laird and ten Kate, 2002).

Biotechnological techniques have revolutionized drug R&D by way of a broadened understanding of molecular triggers, both for good health and disease causation, thereby increasing the precision with which companies identify targets at which to aim drug R&D, and the selection of compounds for research. They also help speed bioassays, and recombinant DNA technologies have enhanced the potential of producing pharmaceutically useful proteins by (a) overcoming the problem with source availability; (b) overcoming product safety issues; (c) providing alternate sources to extract useful proteins, and (d) providing ways to generate engineered therapeutic proteins that display clinical advantages over the native protein (Walsh, 2003: 5).

At the same time, a striking feature of biotechnological developments has been the dual trend in the organization of the drug industry – consolidation on the one hand, and the development of high knowledge/ science-intensive small and medium-sized enterprises (SMEs) or dedicated biotechnology firms (DBFs) on the other. The emergence of the SMEs and DBFs is mainly attributable to the fact that biotechnology remained a mainstay of university laboratories for a long time after its initial successes in the 1970s. The breakthrough discoveries and techniques in biotechnology that established its role as a major drug R&D tool were made in university research environments, while many of the large pharmaceutical firms remained on the periphery (see Powell, 1996). Others, such as Roche, SmithKline, and Merck used recombinant DNA techniques in-house as early as in the late 1970s and early 1980s. The mid-1980s saw the emergence of collaborations between large pharmaceutical firms with smaller biotechnology start-ups to use biotechnological process modifications to produce chemicals of higher value, especially in the area of antibiotics research. A prime example of this is the work of the Ajinomoto company on the production of amino acids.

This generated the need to establish inter-organizational collaborations between large pharmaceutical firms, which historically conducted all their research in-house, and smaller science-intensive enterprises. These companies pursue niche markets by focusing in specific services in the drug R&D chain, and play the role of bridging the gap between the scientific community and the industry.

The SMFs and DBFs have innovative, leading-edge technological skills, are highly competitive amongst each other in offering biotechnology-based services, but either lack the higher expertise or the funds to get involved in the fully-fledged development of products. The pharmaceutical industry, on the other hand, requires these services and has the right risk attitude, financial means, and marketing wherewithal necessary to indulge in the highly uncertain market for R&D based on genetic resources. This is the reason for the increased outsourcing of R&D by the bigger drug companies through allowances, collaborations, and joint ventures with smaller firms.

Although one would normally imagine that the SMEs and DBFs move through different stages of growth – creation, frontier development of technologies, and product development – until now, industry structure shows a clear diversification of labour. Whereas the smaller firms have the scientific expertise, more often than not they do not have the financial means and/or organizational expertise to deal with long drawn-out product development programmes and the risks they entail. The firms thrive and interact with one another based on their functional roles in the R&D process. This corroborates the observation that the reliance on "interorganizational collaborations in the biotechnology industry reflects a fundamental and pervasive concern with access to knowledge" (Powell et al., 1996: 116). The functional roles for the SMEs and the DBFs in the drug industry can include that of independent innovators, problem-solvers using biotechnology methods for specific areas, or niche operators (technological services either through contracts or collaborations).[6]

Biotech firms are immensely competitive, have seen more ups and downs in the past decade, and are more vulnerable to product failures and financial risks. The top ten biotech companies worldwide hold 84 per cent of the total market share in comparison to the top ten pharmaceutical companies which hold 51 per cent (IMS, 2004b). Some of the well-known examples of DBFs that started out as independent innovators are Biogen, Amgen, Biotech, Cetus, and Genetech. Of these, Amgen has been very successful, whereas Chiron and Cetus are effectively non-existent today, and Genetech is wholly owned by Roche. More recently, Raven Biotechnologies, Astex Technologies, and Memory Pharmaceuticals possess enormous potential in independent innovation in the field of cancer, Alzheimer's, nutrition, and animal husbandry (*Time*, 13 December 2004). Good examples of firms that are problem-solvers using biotechnological methods are Allied Signal, Elan and KV Pharmaceuticals. Institutes which specialize in the development of new therapeutic systems for new and established drugs (Ballance et al., 1992) also fall into this category. Of these, Elan is a very interesting example of success, going from a capitalization of over US$2 billion in the late 1990s to severe losses in 2002 and is now on the rebound. Allied Signal, often cited as an example of success in this category, has failed. Niche operators are biotechnology-based firms/institutes offering products and services to the pharmaceutical industry at each stage of the R&D process. They specialize in specific niches in the R&D process, such as *in vitro* and *in vivo* screening activities, bio-assaying and elucidation of chemical compounds, product registrations, and so on (ibid.).

R&D processes for pharmaceutical and botanical medicines

The discussion on R&D processes for pharmaceutical drugs and botanical medicines demonstrates several contrasting issues of relevance to bioprospecting: whereas pharmaceutical R&D is of a much more complex and risky nature than R&D in botanical medicines, the use of raw genetic materials and traditional knowledge as inputs is greater in the botanical medicine industry than in the pharmaceutical industry as it stands today.

The process of pharmaceutical R&D based on genetic resources

Pharmaceutical R&D is a long, drawn-out process that consists of four main stages, starting with the screening process and ending with the development and marketing of the final product. In the pre-screening stage specific search techniques are used whereby genetic resources are selected for further research.

Acquisition of genetic resources and search techniques

Genetic resources can be selected for R&D purposes either through random selection or by targeted techniques. As opposed to random selection, targeted techniques rely on prior information being available on genetic resources for the study of pharmacological activity. There are three main kinds of targeted selection techniques: phylogenetic surveys, ecological surveys, and ethnobotanical surveys (Cox, 1990: 42). Phylogenetic surveys target close relatives of plants that are known to produce useful compounds in the hope of discovering either new homologues or the same compounds of interest that were produced by their close relatives but in a higher concentration. In ecological surveys, plants living in special habitats or possessing certain intrinsic traits are chosen for studying. For example, if there are one or two shrubs surviving amidst large tree species in a dense forest area, then it is assumed botanically that the shrubs possess some special survival characteristics. These shrubs are then collected and tested for their special characteristics. In ethnobotanical surveys, plants used by indigenous peoples in their traditional medicine are chosen for study, since such plants are thought to possess medicinal compounds of interest (ibid).

Drug discovery and pre-clinical development

Selected samples are then dried and extracts run through a range of screens that are designed to detect important chemical/therapeutic activity in the samples. Where useful biological activity is detected in a sample, it is described as a positive "hit" (ten Kate and Laird, 1999). The hit rates among screening institutions can vary widely depending upon the number and types of screens that are available to them and the methods they employ in detecting biological activity (Rosenthal, 1998). After such preliminary screening processes (such as hippocratic screens and brine shrimp lethality assays), the samples that show therapeutic activity of interest are re-tested using secondary screening methods. In secondary screens, samples are tested against the target of the R&D programme using *in vitro* screens or mechanism-based screens. *In vitro* models employ cell lines representative of the disease condition grown in an artificial medium. In contrast, mechanism-based screens contain protein targets that detect samples which may operate by the anticipated mechanism of action relevant to the disease. When the sample shows activity in the *in vitro* process or mechanism screen, the next step is to isolate a pure chemical from the extract, which is a complex mixture of natural products. At this stage, the most important task is to ensure that the sample is active enough and not toxic.

This process of primary screening, followed by secondary screening and isolating the "leads" forms the basis of drug discovery and research.

Pre-clinical drug development starts with the isolation and structural elucidation of the potential therapeutic agent found in the sample. Isolation and elucidation of natural products is a very complicated task that may involve between 40 and 50 steps before researchers arrive at the active biological compound or compounds within a crude natural extract that are responsible for the therapeutic effect. Often, due to the complexity inherent in this process, if any of the molecules (which may be semi-synthetic or synthetic) in one of the intermediate stages show sufficiently good activity against the targets of the programme, they are considered for drug development purposes. The agents that show significant *in vivo* activity in appropriate animal models enter pre-clinical and clinical development. The key steps involved in the preclinical development process are:

(a) Development of an adequate supply of the agent to permit pre-clinical and clinical development;
(b) Formulation studies to develop a suitable vehicle to solubilize the drug for administration to patients, generally by intravenous injection or infusion in the case of cancer;
(c) Pharmacological evaluation to determine the best route and schedule of administration to achieve optimal activity of the drug in animal models, the half-lives and bio-availability of the drug in the body after administration, the rates of clearance and the routes of excretion, and the identity and rates of formation of possible metabolites (that would limit their availability to treat the required disease condition); and,
(d) Investigational New Drug (IND)-directed toxicological studies to de-

termine the type and degree of major toxicities in rodent and dog models. Toxicity tests look for any possible side effects of the drug candidate. Chemical modifications may be necessary either to reduce toxicity levels or potential side effects of the drug candidate or even to increase its degree of activity. These studies help to establish the safe starting doses for administration to human patients in clinical trials (Cragg, personal communication, 2004).

There is usually a requirement for a two-year toxicity study to be conducted in two different species for all drugs, except those for cancer and HIV, thus the time-frames are affected by the disease(s) that the drug candidate is targeted against. Furthermore, all toxicity testing and subsequent clinical work must be performed with the compound isolated and purified to the same level of defined impurities under current good manufacturing practices (cGMP) conditions. Although this is important when dealing with all drug candidates, it is critical in the case of a naturally sourced drug because, even before any commitment to clinical studies is made, there must be sufficient chemically defined compound available for analysis to complete the work.

If a drug candidate successfully advances through these four stages, a patent application may be filed and an application submitted to the relevant governmental authority for clinical development.

Clinical development

Clinical development consists of several phases. Phase I is designed to demonstrate a lack of toxicity in normal patients (or any toxicity that is at such low levels that it is tolerable when compared to the benefit from such an agent). Phase II (and some intermediate Phase I studies) are then used with patients suffering from the disease and the activities seen compared against a placebo treatment (this is a "double-blinded study", as neither the patient nor the physician knows who is getting the drug candidate and who is not). The intermediate Phase I studies in volunteers are designed to determine the pharmacokinetic and pharmacodynamic properties of the drug candidate.[7] Provided no significant toxicities are seen and there is a demonstration of efficacy, then a series of much larger Phase III trials are begun, again in a double-blinded mode. If these are successful, then an application is made for approval as listed below. These phases can take up to five to six years to complete.

Efficacy and drug marketing

The data on the drug's performance in the clinical trials demonstrates its efficacy and forms the basis of an NDA application to regulatory authorities. If the trials end successfully, the drug is filed for approval by the appropriate body (in the USA, this is the FDA, and in Europe it is the

European Agency for the Evaluation of Medicinal Products (EMEA); only after this is it allowed to be marketed. This can be done either through the firm's own marketing department (which is the case with many large pharmaceutical firms) or by licensing the production of the drug (see ten Kate and Laird, 1999: 54–55).

Drug R&D for botanical medicines[8]

In contrast to the long processes of R&D required for pharmaceutical drugs, botanical medicines are relatively simpler preparations.

The first stage involves the gathering of information on the useful properties of plants. More often than not, traditional knowledge plays a very important role in the development of botanical medicines, not only by pointing to useful properties of plants, but also because in many cases the entire botanical product may be based on the traditional usage itself. Traditional knowledge can also play a major role in proving the safety and efficacy of botanical medicines. The selected plant is then tested to check if the relevant activity can be found in any chemical or chemical group present in it, or if its known chemicals possess certain pharmacological properties. If this turns out to be the case and the plant does contain pharmacological properties of commercial interest, development of the botanical medicine will require the availability of large quantities of raw plant material.

At the next stage, the processing companies test plant material for contamination (e.g. pesticides). This stage is also usually undertaken by intermediaries, for example, brokers who specialize in preparation of extracts. In the final stage, the processed ingredients are supplied to the companies who manufacture the finished products (by fixing the form and the consistency according to in-house formulae).

Bioprospecting and drug research

Bioprospecting – either for genetic resources alone or traditional knowledge, or both – has been a source of many useful drugs even historically. Examples of useful drugs produced from natural products that were also in use in traditional medicine are *Catharanthus Roseus* (Rosy Periwinkle, generating an annual sales of US$200 billion annually in 1995), *Taxus Brevifolia* (Himalayan Yew, annual sales of US$941 million in 1997), *Rauwolfia serpentina* (Indian Snakeroot, estimated annual sales of US$260 million in 1994), amongst many others (ten Kate and Laird, 1999; Fellows and Scofield, 1995), though often the use of the isolated drug does not correspond to the original use of the species in traditional medicine. Bioprospecting still represents an important activity for the industry given the continued interest in natural products as a source of new drugs.

Genetic resources for the pharmaceutical and botanical industry

Generally, in the case of pharmaceutical R&D based on genetic resources, there are several stages that require raw genetic material. Whereas initial screening and isolation of lead compounds only requires around 5 kilograms of dried material, confirmatory screens and initial development requires around 50 kilograms of dried material (McChesney, 1992, cited in Aylward, 1995: 114). But pre-clinical and clinical testing and large-scale production of the product require much larger quantities of the genetic material (Cragg, Newman, and Snader, 1997). It has been estimated that pre-clinical and clinical testing may require around 200 tons of dried material whereas mass production may require as much as 200,000 tons per year (McChesney, 1992, cited in Aylward, 1995: 114). Such high quantities, though, may only hold good for the first time the drug is produced as subsequently attempts are usually made to find ways of producing high-yielding cultivars or intermediates which are fast-growing versions of the original species (as in the case of Taxol), thus providing renewable resources.

Furthermore, not all products sourced by genetic resources require mass cultivation for production of the final drug. Of over a 100 prescription drugs available today, many are produced by direct chemical synthesis or semi-synthesis (Walsh, 2003: 53). Of the several categories identified earlier on in this chapter, biologicals and natural product-based drugs are usually produced naturally – that is, they require inputs of the genetic resource in question. In the case of drugs that are derived from natural products, there are cases where the raw material is still required for the production of the drug. Often, natural products are extremely complex structures and thus it is likely that laboratory synthesis is not possible, at least not in an economically feasible way. A good example of this is the anti-cancer drug Topotecan. Although Topotecan is a synthetic derivative of the natural product camptothecin that is found in a Chinese plant, the production of the drug still begins with camptothecin (see ten Kate and Wells, 1998).

The botanical drug industry has also witnessed a shift from simple preparations that involved minimal processing, towards more complicated herb combinations in medicines and phytomedicines drawn from traditional medicines worldwide. Many of the plants used in botanical medicines come from all over the world. Whereas Asian species play an increasing role in botanical medicines due to the importance of traditional Chinese medicine, African and Latin American species are more limited (Laird and ten Kate, 2002: 258). Botanical medicines, as they stand today, are mostly comprised of total plant extracts (ten Kate and Laird, 1999: 80). This underscores the importance of raw genetic material supply in this trade and the role of bioprospecting in sourcing them. According to a study by Lange (1997), of the plants used in botanical medicines sold in Germany, only 16 species occur in Europe; 63 were found to be from Africa, 90 from tropical Asia, 8 from Australia and New Zealand, and 248 species from temperate Asia (Lange, cited in ten Kate and Laird, 1999: 83).

Traditional knowledge for pharmaceutical and botanical R&D

It is well accepted by now that many of the plant-derived drugs we have today are to be attributed to the input of traditional knowledge (Taylor, 1965; Farnsworth, 1985; Lewis, 1992, among others). Literature is replete with studies that elaborate upon the contributions of traditional medicine to modern drug research. Traditional knowledge can, more precisely, help to narrow down the search for useful species with bioactive materials. In such cases, ethnobotanical knowledge serves as a starting point of drug development. It has been estimated that by consulting indigenous peoples, bio-prospectors can increase their success ratio by 400 per cent, depending on the disease condition.[9]

In some other instances, traditional knowledge can directly help identify precise uses of genetic resources that are then tested and drugs with those very precise uses are developed. Although fewer, in such cases traditional knowledge contributes the main information that leads to the development of the commercial product.

Some examples of products derived from traditional knowledge but with different uses from the original are *Podophyllum peltatum* (traditional use: cathartic and antilimnitic, clinical use: anti-cancer), *Cloeus forskolii* (traditional use: cardiovascular, clinical use: anti-metastatic) and *Homolanthus nutans* (traditional use: anti-diarrhoeal, clinical use: antiviral) (Cunningham, 1996: 12). Piper futokadsura (traditional and clinical use: broncho-asthma), Cinchona pibescen (traditional and clinical use: anti-malarial) and Artemesia annua (traditional and clinical use: antimalarial) are examples of products that are identical to the original traditional use (ibid.).

In the case of botanical medicines, however, traditional knowledge plays a very important role. Often, botanical medicines are not only based completely on the traditional medicinal use in question, but the efficacy of use in traditional medicine is also used as a selling point to attract customers in foreign markets, as already noted. Some popular examples of botanical medicines that have major markets in the EU and USA are: *Ginkgo biloba*, Ginseng, *Echinacea*, St. John's wort, Aloe, and Golden Seal (see ten Kate and Laird, 1999; Laird and ten Kate, 2002).

Bioprospecting contracts for drug research

Bioprospecting contracts that have been witnessed in the market over the past decade or so show three main trends: (a) long-term collaborations with multilateral contractual relationships between various actors involved in the drug discovery process, or bilateral/trilateral relationships between the drug company, source country and/or community; (b) a series of bilateral contracts; and (c) spot-market transactions. In both, (b) and (c), the tangible genetic material as well as traditional knowledge based-information gains in value as it climbs the R&D chain. The examples of contracts discussed in this section are not exhaustive; they are only of an illustrative nature.

Research collaborations

Most previous and on-going research collaborations are some sort of public–private or public–public forms of partnerships. A good example of research collaborations are the drug research projects underway as part of the International Cooperative Biodiversity Groups (hereafter, ICBGs), a project of the United States Government.

The international cooperative biodiversity groups

The ICBGs, initiated in 1991, endeavour to develop a cooperative programme to meet the intertwined needs of biodiversity use and conservation, ethnobotanical knowledge, and pharmaceutical drug discovery (Timmerman, 1997: 219).[10] Initially, the programme was funded and guided by three US governmental agencies – the National Institute of Health (NIH), the National Science Foundation (NSF) and the US Agency for International Development (USAID) (Rosenthal, 1997: 254), but USAID withdrew later, and the programme is currently funded by the NIH, the NSF and the US Department of Agriculture (USDA).[11] The contractual structures of the ICBGs are meant to serve as innovative examples of the kinds of contracts that can be struck to balance intellectual property rights with other needs between partners for drug discovery and development, and, simultaneously, to serve the goals of drug discovery, equitable benefit-sharing, and biodiversity conservation (Timmerman, 1997).

For the ICBG programmes, appropriate design is supposed to include, as Rosenthal (1997) notes: (a) active participation of source country individuals and organizations from the planning stage onward; (b) multidisciplinary research and diseases of both local and international significance; (c) local training and infrastructure development in both drug discovery and biodiversity management; (d) biodiversity inventory and monitoring; and, finally, (e) equitable intellectual property and benefit-sharing.[12] In this spirit, the different ICBGs display the various possibilities of contractual structures (ranging from advance payments to royalties to scientific capacity building arrangements in the source countries) and protection mechanisms for ethnobotanical knowledge. Several different ICBG programmes have been initiated as part of this initiative since 1992, some of which are discussed briefly here. Some others, such as the ICBG Peru and the ICBG Panama have not been renewed, whereas others such as the ICBG West Africa are still ongoing.

(a) Biodiversity utilization in Madagascar

This ICBG is led by Dr. David Kingstone of Virginia Polytechnic Institute and State University (Virginia Tech) working in conjunction with Conservation International (CI) of Suriname, Missouri Botanical Gardens, Bristol-Myers Squibb Pharmaceutical Research Institute (BMS), the National Herbarium of Suriname, and the Bedrijf Geneesmiddelen Voorziening Suriname (BGVS).

(b) ICBG Peru

The contractual structure of the Peruvian ICBG consisted of a number of bilateral arrangements between all the participants. These were: Washington University, Monsanto-Searle Pharmaceutical Company, the Aguaruna peoples of Peru, Universidad San Marcos, and Universidad Peruaua Cayetano-Heredia. The agreement between the Aguaruna community and Monsanto-Searle to disclose ethnobotanical knowledge was guarded with the help of a know-how licence that caters to issues of sharing knowledge as well as benefit-sharing with the communities (Tobin, 2002). There was a Biological Collecting Agreement between the Aguaruna Peoples (represented by three local indigenous organizations and Washington University for collection of tangible samples for screening as well as higher research collaboration and benefit-sharing (Rosenthal, 1997).

The collaboration between the native Aguaruna peoples of Peru and the research team from Washington University, led by Dr. Lewis, has had initial success in identifying effective samples for some diseases (including several types of cancer) via bio-assay techniques (Lazaroff, 2000).

(c) Ecologically guided bioprospecting in Panama

A five-year grant was given to the ICBG in Panama that sought to link drug discovery from rainforest plants to biodiversity conservation. This ICBG employed the ecological method to target plants that could have useful medical compounds. The diseases targeted by this programme are: HIV, malaria, and various forms of cancers. The aim of biodiversity conservation is sought to be achieved through a local foundation, Fundación Natura, by creating an environmental trust fund that will receive the largest shares of the revenues associated with the discovery of a novel medicine.

(d) Costa Rican conservation and ICBG

A first five-year grant was given to Cornell University in cooperation with INBio of Costa Rica and Bristol Myers Squibb to focus on a programme targeting on tropical insects and other species as potential sources of new drugs. INBio not only trains Costa Rican scientists to conduct field studies and drug discovery studies, but its scientists prepare extracts from biological materials and also carry out some screening programmes as part of their contribution to this project. Bristol Myers Squibb is screening the extracts prepared at INBio.

(e) ICBG West Africa

This project was initiated to explore the rich diversity of moist tropical forests in Africa. The programme has now been extended for a second five-year term and the partners are: the Walter Reed Army Institute of Research in cooperation with the University of Yaounde, Cameroon; the Bioresources Development and Conservation Programme, Nigeria; the University of Jos, Nigeria; the International Centre for Ethnomedicine and Drug Development, Nsukka, Nigeria; the Smithsonian Institution; and the universities of Utah, New York, and Minnesota.

(f) ICBG Drug Discovery Programme among the Mayans of Mexico

This programme, led by Brent O. Berlin of the University of Georgia in Athens in collaboration with the scientists at the College of the Southern Frontiers in Chiapas, Mexico, and Molecular Nature Ltd., aimed at evaluating the pharmacologically important tropical plants and fungi utilized by the Mayan speaking communities of Southern Mexico. Due to contractual problems, the project was discontinued in 2001 (see Chapter 4 for a detailed discussion).

The collaborative arrangements of the National Cancer Institute, USA

Another interesting set of collaborative arrangements can be seen in the work of the National Cancer Institute (NCI) of the USA. The NCI has carried out investigations of collections of plants and marine organisms in over 25 countries. These have been made possible through contracts with botanical and marine biological organizations working in close collaboration with qualified source country organizations. The recognition of the value of the natural resources (plant, marine, and microbial) being investigated by the NCI, and the significant contributions being made by source country scientists in aiding the performance of the NCI collection programmes, have led the NCI formulating its Letter of Collection which specifies policies aimed at facilitating collaboration with, and compensation of, countries participating in its drug discovery programme (Cragg and Newman, personal communication, 2004). One of the NCI agreements is discussed here.

(a) Potential anti-HIV agents from *Calophyllum* species, Sarawak

What began with an organic extract of leaves and twigs of the tree, *Calophyllum lanigerum* collected in Sarawak in 1988, led to the development of calanolide A and calanolide B. Since both were found to show significant anti-HIV activity, the NCI patented both calanolides in 1995 and awarded the licence for their exclusive manufacturing to Medichem Research, Inc. The agreement required Medichem to negotiate an arrangement with the State Government of Sarawak for collection of latex from the tree for conducting R&D. By 1996, the State Government of Sarawak and Medichem had formed a joint venture company called Sarawak Medichem Pharmaceuticals Incorporated, which has sponsored Phase I clinical trials. Presently, Phase II clinical trails with patients infected by HIV-1 are in progress (Cragg and Newman, personal communication, 2004; also see ten Kate and Wells, 1998).

Private research collaborations

There are very few cases where large pharmaceutical firms in fact approach developing countries directly for access to genetic resources and/or traditional knowledge. Most private research collaborations take place in the stages of intermediation discussed in the next section, where smaller firms/research institutes or botanical gardens offer important services or through spot-market transactions.

National agencies can also play the role of such intermediaries or even replace them in certain circumstances, if they perform the same task of adding value. A good example of this is INBio, which was established by the Costa Rican government. Another is the UZACHI, which is a coalition of communities of Zapotec and Chinantec ethnicity in the Oaxaca region of Mexico. Both these agreements are discussed here.

(a) Merck INBio

The Merck–INBio agreement has often been cited as one of the most promising agreements in bioprospecting and as one of the best examples of what can be called a "win-win" situation. INBio is a private non-profit institution that was established in 1989 by the Costa Rican government to conserve its biodiversity through facilitating and stimulating its use in a sustainable way. In 1991, Merck, one of the biggest pharmaceutical concerns in the world, and INBio entered into an agreement whereby INBio agreed to provide Merck with plant, insect, and environmental samples for an upfront payment of US$1.185 million, and Merck also agreed to pay royalties on any products developed therefrom (ten Kate and Wells, 2001). Additionally, INBio scientists received technical training from Merck. This agreement, as mentioned in Chapter 1, concluded in 2001 after Merck made a last grant of US$130,000 to INBio (Dalton, 2004b: 599). The results of the programme on any potential drug candidate are unclear.

(b) Sandoz AG of Switzerland (now Novartis) and UZACHI[13]

The private bioprospecting contract between UZACHI and Sandoz AG (now Novratis) has been noted for its sustainable aspects (Chapela, 1997). The contract for bioprospecting contains an upfront payment, royalties (in terms of milestone payments), and provisions of training and capacity building. The agreement for initial collection started in 1995 and ended in 1998, where Novartis mainly focused on fungal resources of the UZACHI. Collections were part of the Biolead Project of Novartis. There was no mention of traditional knowledge in the agreement itself in accordance with the wishes of UZACHI, but the conservation of biodiversity initiated by the UZACHI using the funds generated by the agreement involved traditional knowledge.

The success of this agreement in terms of benefit-sharing for the communities and the channelling of these benefits into *in situ* conservation is to be attributed to the fact that the communities had a integrated conservation-related management plan at the inception of the agreement that Novartis had to merely support through its revenues. For the UZACHI, the main achievement of the project until now has been the realization that their fungal resources are very valuable. The UZACHI have been successful in channelling the limited funds (in terms of upfront payment) and the capacity building exchange that took place between the two parties to establish the Mycological Facility, Oaxaca (hereafter, the MFO). Through the MFO, the community has developed their own cadre of technicians and also hired professional people from other Mexican cities. But these outsiders are not constant in the MFO; the UZACHI maintain the facility completely. The MFO has been running now for 15 years and is completely supported through UZACHI funds that come from other activities of the communities, such as forest produce. The communities use the MFO not only to explore the commercial potential of their resources but also to research and enhance their *in situ* management strategies. In 1997, using the funds generated through the Novartis agreement, the UZACHI established a DNA capacity for genetic analysis of crop diversity in the MFO.

Bilateral contracts/spot-market transactions

It is in the case of plant genetic resources that raw genetic material is required for pharmaceutical R&D or the preparation of botanical medicines. Most of such genetic material is sourced through either bilateral contracts or spot-market transactions. The 1980s and particularly the 1990s saw the rise of several kinds of intermediaries in the bioprospecting market. Apart from smaller firms that try to fill a niche in operations in the drug R&D process, as identified earlier, there are also firms, university research institutes, or even private research institutes such as botanical gardens that have widened their activities to fill the gaps in the supply of genetic samples.

A good example of a DBF that performs the role of coordinating biotechnology-related services is the UK-based biotechnological firm Biotics Ltd. Biotics is a sub-contractor which sub-contracts the collection of plant samples to developing countries and carries out the initial processing and the extraction of the samples in its British-based laboratories, and then enters into contracts with pharmaceutical firms to whom it supplies the extracts (Aylward, 1995). The American firm Rain-Tree Nutritions, Inc. offers services for conducting intermediary research and the harvesting of rainforest plants for pharmaceutical use. Indena, an Italian company, grows a species of Taxus (European Yew) from which they isolate a precursor (A-Baccatin), which is closely related to Taxol, and which they then supply to BMS for conversion into Taxol (Cragg, personal communication, 2004).

As these examples show, more commonly, the process of drug discovery based on genetic resources comprises a series of contractual arrangements for the transfer of tangible genetic resources and intangible information related to them.

In the case of botanical medicines, raw material is traded in bulk (Laird and Guillen, 2002). In this sector, too, a large network of intermediaries help source the plants. These can either be supply companies (cultivators or wild crafters of raw material, and includes wholesalers, exporters, traders, brokers, agents, and bulk processing companies), manufacturing and marketing companies (including branding and labelling companies) and consumer sales counterparts (brokers and distributors of finished products) (Laird and ten Kate, 2002: 259). These intermediaries may either cultivate the material themselves, or purchase in the market through "spot market" traders. Supplying such raw genetic material is a large occupation developing especially amongst local, subsistence-based people in developing countries. These "suppliers" trade the material at spot market prices that vary according to market demand. Although several examples of sudden rises in price of bulk material exist (e.g. Kava and Cats Claw), it needs to be cautioned that in many cases, such price rises may only be a result of market hype and have little to do with real increases in global demand.

These "spot market" dealings in the exchange of resources for bioprospecting also extend to traditional knowledge-related information. Reports abound of pharmaceutical companies dealing with individual collectors who come to them with plants that are said to have ethnobotanical significance (Laird, 1993: 112; Cunningham, 1996).

Notes

1. The author is very grateful to Gordon Cragg and David Newman of the Natural Products Branch, National Cancer Institute, USA, for extensive comments and suggestions on earlier drafts of this chapter.
2. In their calculation, Newman et al. (2003) use a category called "natural product mimic", which refers to those drugs which are synthetic but are modelled on a natural product inhibitor of the molecular target of interest or mimic it.
3. Tufts Center for the Study of Drug Development (2001) estimate, cited in Dutfield (2003: 91).
4. For a detailed discussion on how much the pharmaceutical industry really spends on R&D, including a critical evaluation of the Tufts Center estimate of US$802 million, see Angell, M (2004), Chapter 3.
5. Similar estimates are presented by McChesney, 1996. Halliday et al. (1992) estimate that on an average, only one new product emerges from 3,645 compounds synthesized annually.
6. For a slightly different categorization, see Balance et al. (1992) or Saviotti (1998).
7. Pharmacokinetic properties are those that determine the fate of the drug within the human body, such as: absorption into the body, distribution within the body, metabolism by the body and excretion from the body. As a result these studies are useful to indicate not only the appropriate dosage but also the precise route of administration. Pharmacodynamic properties of a drug refer to the way it interacts with the cells/organs it is meant to act upon, and any side effects it may generate (Walsh, 2003: 70).
8. The information in this sub-section is based on ten Kate, K. and Laird, S.A. (1999: 84–85).
9. According to Reid et al. (1993: 150), given the need to screen roughly 10,000 chemicals to find a single lead, there is a one-in-four chance of a lead being developed into a commercial product, a 5 per cent discount rate, a 10-year wait before a product is ready to be marketed, and 15 years of patent protection while it is being marketed, and, on the

assumption that a drug, if discovered, generates US$10 million net annual revenues, the present value of the agreement to the supplier is only US$52,200. Therefore, there is a 97.5 per cent chance that the 10,000 chemicals will not result in any commercial product at all. Contrasting this with natural products, they conclude that "the prospects for success are raised, since any extract from a species will contain hundreds or thousands of different chemicals that may result in a pharmaceutical 'lead'. Thus, the probability of success could easily be ten times that of the example above, thus producing promising leads of about 1 per 1000 samples. The probability of developing at least one commercial product in the example would then rise from 2.5% to 22%."

10. Also see Grifo (1996), Grifo and Downes (1996), Rosenthal (1997) among others.
11. See http://www.fic.nih.gov/programs/icbg.html for more recent developments.
12. Rosenthal (1997: 255).
13. This section is based on Gehl Sampath and Tarasofsky (2002).

3

International policy dimensions of bioprospecting

Introduction

If user countries want efficacious access to biological resources and to traditional medicinal knowledge, and if source countries want revenues in return, why do we not see consensus developing on these issues? What constitutes a country's right to regulate access to genetic resources? Is this a blanket right or can it be denied only on grounds of sustainable use and conservation of biodiversity? How and on the basis of which factors can source countries set terms of compensation for use of genetic resources? Under which circumstances can such terms and conditions lead to reinforcing drug R&D based on genetic resources and under which circumstances can they potentially stifle R&D efforts altogether? Can conventional intellectual property rights (IPRs) protect knowledge holdings of a different kind, namely, traditional medicinal knowledge? Under which circumstances should traditional medicinal knowledge of relevance to drug R&D be protected? How can contracts on traditional medicinal knowledge be designed and what sort of rules and institutions are required to facilitate them? What factors are critical in determining "fair and equitable" compensation for such knowledge?

To answer at least some of these questions satisfactorily, this chapter conducts a detailed analysis of the interaction between the TRIPS Agreement and the CBD to pin down the main spheres of interaction between the agreements for use of genetic resources and traditional medicinal knowledge in drug R&D. Section I deals with the provisions of the CBD that are of utmost relevance to bioprospecting. Section II is a similar investigation into the TRIPS Agreement. Section III analyses the link between intellectual property rights, biotechnology, and biodiversity. It explains the main areas of interface between the TRIPS and the CBD that are relevant for bioprospecting. An investigation into the politics that divide the technology-rich North and the biodiversity-rich South on this topic shows that most problems in reaching consensus on the unresolved issues are due to varying political interests and lack of legal clarity as to the use of certain terms in both agreements. Section IV revisits Article 15 and Article 8(j) to show that when the rights to access and traditional knowledge are considered along with the limitations that the CBD itself places on them, the main legal controversies regarding their scope and content can be resolved to a large extent.

The Convention on Biological Diversity, 1993

Although the long-term availability of biological resources to all users depends on its sustainable use and conservation, no one user group has the perfect incentives to internalize the externalities it generates in biodiversity use (OECD, 1995). The CBD reflects the wider international acceptance of this fact. It also embodies the recognition that national governments have a primary role to play in ensuring sustainable biodiversity regimes within which commerce, resource management and environmental institutions can act in tandem to prevent destruction, extinction, or alienation of biological resources (WRI et al., 1992).[1]

There are several reasons for which the CBD, which came into force in 1993 with 172 countries, should be lauded. Apart from being an internationally binding agreement with the largest ratification status, it establishes a review mechanism (the Conference of the Parties) through which the signatories continue to meet to review progress and act on biodiversity-related issues every two years. But the Convention's most important point of departure from earlier international environmental agreements is the recognition that conservation and sustainable use of biodiversity can only be tackled when viewed within the economic context in which biodiversity operates (WRI et al., 1992: 28–29). The CBD mirrors the realization that biodiversity conservation is far more complex an issue than nature conservation and therefore entails taking into account a more offensive effort that tackles use and conservation of each component of biodiversity (ibid.: 5). As a result, it envisages a comprehensive cross-sectoral approach to biodiversity conservation that seeks to integrate use with incentives to conserve, from a global to a local level (ibid.: 13). The recognition of property rights is one such set of incentives under the Convention and is aimed at encouraging beneficiaries to continue efforts of conservation and sustainable use of biological resources. The sharing of benefits that arises from commercialization of genetic resources with source countries is another incentive to encourage conservation efforts (Bragdon and Downes, 1998).

The three main objectives of the Convention are spelt out in Article 1, which are:

(a) the conservation of biological diversity;
(b) the sustainable use of its components, and;
(c) the fair and equitable sharing of benefits arising out of the commercialization of biological resources, including by appropriate access to genetic resources and by appropriate transfer of relevant technologies, taking into account all rights to those resources and to technologies, and by appropriate funding.

To reinforce these objectives, the CBD recognizes national sovereignty on genetic resources that lie within a country's territory. This recognition accorded under Article 15 is primarily meant to enable and empower national governments to identify national priorities in biodiversity conservation, to set out legal rules and institutions to govern local actions that affect biodiversity use and to enhance accountability in use and conservation of biodiversity (WRI et al., 1992: 26).

The CBD also identifies a wider range of measures that can be used for the implementation of its objectives. These are contained in Articles 6 to 14 of the Convention and include: (a) identifying and monitoring components of biological diversity, such as specific ecosystems (Art. 7); (b) establishing a system of protected areas (Art. 8); (c) adopting measures for and *ex situ* conservation; (d) integrating genetic resource conservation considerations into national decision-making and adopting incentives for the conservation of biological resources (Art. 10 and 11); and, (e) developing assessment procedures for ensuring that impacts on biological diversity are considered in project design (Art. 14) (see Miller, 1997: 171).

Apart from the identification of such broad-based measures, the CBD does not specifically prioritize the measures nor does it provide any operative details, thus leaving member countries with a lot of flexibility to implement its framework.[2]

Traditional knowledge and its protection under Article 8(j)

Article 8(j) titled "*In situ* Conservation" reads as follows.

Each Contracting Party shall, as far as possible and as appropriate ... subject to national legislation, respect, preserve and maintain knowledge, innovations and practices of indigenous and local communities embodying traditional lifestyles relevant for the conservation and sustainable use of biological diversity and promote their wider application with the approval and involvement of the holders of such knowledge, innovations and practices and encourage the equitable sharing of the benefits arising from the utilization of such knowledge, innovation and practices.

This right falls within the CBD scheme of granting incentives at the local level for sustainable use and conserve *in situ* biodiversity and can therefore be called an "indirect incentive".[3] In this instance, the CBD suggests that knowledge, innovations and practices of indigenous and local communities are useful for biodiversity use and conservation and that they have a right over it, subject to national legislation (Dutfield, personal communication, 2002).

Article 8(j) is broad and all encompassing in nature. It not only includes the "knowledge, innovations and practices of indigenous and local communities" but also mentions them in conjunction with conservation and sustainable use. The expectation is that national governments not only recognize the right, but also provide for participatory mechanisms for the exercise of this right so that communities can share in the benefits arising out of commercialization of their resources. This is evident from the second part of the Article that asks State Parties to "promote their wider application with the approval and involvement of the holders of such knowledge, innovations and practices".

Article 8(j) also marks a watershed in the discourse on rights of indigenous and local communities; a very intensely debated theme in international environmental law ever since Convention 169 of the International Labour Organization recognized their right to self-determination.[4] However, this Convention was never ratified, it ended up being a mere vantage point for controversial discourses on the rights of indigenous and local communities (see Posey, 1996). Post Article 8(j), legal scholars have expressed the possibility of ushering in national laws based on this Article, which also covers other rights of communities, such as their right to life, livelihood, etc.[5]

Article 15 and the right to regulate access to genetic resources

Biodiversity and its three major components, namely, species, ecosystems and genes, possess different values for use, both from an individual and social perspective. These can be classified broadly into direct and indirect use values. *Direct use values* consist of those parts of biodiversity which can be appropriated as goods and can be traded in the market

(Mountford and Keppler, 1999: 135–136). Genetic resources are a subcomponent of biodiversity that largely comprise of direct use values – hence they are directly tradable in markets, although they still possess certain indirect values that cannot be fully captured by private property rights (Lassere, 2002). In contrast, indirect use consists of values of biodiversity that cannot be easily valued in monetary terms. There are several indirect use values of biodiversity. One of these, called ecosystem services, comprises all those functions of a given environment that contribute to the creation of direct values, like aspects that control floods, droughts, soil erosions, etc. (OECD, 1999: 29). Another indirect use value of biodiversity is the option value or quasi-option value of biodiversity, a concept that has been discussed in detail in Chapter 6.

The CBD, through Article 15 establishes property rights on genetic resources, the main tradable component of biodiversity, thereby converting biological diversity from an open-access resource at the international level to a common property resource at the national level.

Article 15(1) reads as follows:

> Recognizing the sovereign rights of States over their natural resources, the authority to determine access to genetic resources rests with the national governments and is subject to national legislation.

The main purpose of recognizing the right of countries to regulate access to genetic resources seems to be based on the need to ensure that local needs and priorities form an integral part of biodiversity policies. Accordingly, each Member Country is obliged to take legislative, administrative and policy measures to implement this provision at the national level. Some other sub-provisions of Article 15 provide the chief support mechanisms on the basis of which national rules of access should be designed. Article 15(2) talks of *facilitatory access*, Article 15(4) of *mutually agreed terms* and Article 15(5) enumerates the concept of *Prior Informed Consent*. The concept of prior informed consent strengthens the right of the source country to control its genetic resources, by stipulating that access should only take place with prior informed consent of the source country.[6] Most legal literature views Prior Informed Consent (hereafter, PIC) as one of the chief legal mechanisms to support the authority of source countries to grant *valid* access.[7] As Mann (1996: 28) notes, "The PIC process is widely recognized in the literature to provide a sound procedural footing to the bargaining power needed by developing countries to balance the resources and skills of biotechnological companies. Through it, bilateral agreements can be reached that include the full range of possible technology and financial awards."

Article 15(4) on mutually agreed terms (hereafter, MAT) makes contractual bargaining a pre-condition for access (Straus, 1998: 104).[8] Both PIC and MAT requirements are complementary to one another – whereas the emphasis of PIC is on the provision of legal rules for access that lay down the terms and conditions of use and transfer, the focus of MAT is on the role of access institutions as facilitators of contracts.

Although details of how these mechanisms are to function have been left open for countries to decide, one guiding element is Article 15(2) that underlines the obligation of source countries to facilitate access to genetic resources.[9]

The concept of benefit-sharing

The CBD embodies the concept of benefit-sharing in both a general and specific way. Article 1 generally stipulates "fair and equitable sharing of benefits" as an incentive for right holders and stakeholders to continue their conservation and sustainable use efforts. Specific guidelines on benefit-sharing are contained in several Articles, such as Article 16 that talks of technology transfer as a form of benefit-sharing or Article 19 that talks of a fair handling of benefits of biotechnology research between the developed and developing countries (see Glowka et al., 1994). In this more specific sense of benefit-sharing, the Convention gives developing countries a chance, if utilized smartly, to use their genetic wealth to propel their way into the biotechnology era.

But the Convention's mandate on benefit-sharing, both general and specific, has big gaps in it. First of all, although the right of source countries to be compensated for provision of genetic resources is envisaged as one way to mobilize resources for the protection of biodiversity, the Convention does not codify the relationship between the generation of revenues through such activities and biodiversity conservation *in a direct way*. Secondly, the Convention does not define "appropriate" compensation for the use of genetic resources. Lastly, although it mentions specific forms of benefit-sharing, it does not provide guidelines as to which form of benefits should be shared under what circumstances, again leaving a lot of room for countries to deliberate amongst themselves.

Although there is a broad consensus in literature that one of the primary mechanisms to negotiate benefit-sharing is the PIC and MAT process, interpretations of benefit-sharing are varied. Benefit-sharing can range from royalties, to upfront payments, rewards, and to transfer of biotechnology.[10] As a result of this ambiguity, scholars have also expressed the fear that the PIC process and benefit-sharing provisions may end up being a rent-seeking process where developing countries unduly benefit from those wishing to use their genetic resources (see, for example, Straus, 1993: 607).

The Agreement on Trade Related Aspects of Intellectual Property Rights, 1995

Since information that is once produced cannot be controlled effectively by the creator, there is a lack of incentives for people to invest in its production. Intellectual property protection is a mechanism that seeks to encourage the production of useful information in society by granting temporary monopoly rights to inventors on their creations. In recent times, intellectual property protection has become an important market-leveraging instrument in global trade, due to the growing research and the capital-intensive nature of new technologies. The TRIPS Agreement, which is one of the annex agreements to the Agreement that established the World Trade Organization in 1995, underscores the important role that intellectual property protection plays in global trade today.

General principles of the TRIPS Agreement

The TRIPS Agreement attempts "to reduce distortions and impediments to international trade, and taking into account the need to promote effective and adequate protection of intellectual property rights, and to ensure that measures and procedures to enforce intellectual property rights do not themselves become barriers to legitimate trade" (Preamble). The objective of the Agreement is to ensure that the protection and enforcement of intellectual property rights contributes "to the promotion of technological innovation and to the transfer and dissemination of technology, to the mutual advantage of producers and users of technological knowledge and in a manner conducive to social and economic welfare, and to a balance of rights and obligations" (Article 7).

To achieve this, the Agreement provides for new rules, among others, concerning:

(a) adequate standards and principles concerning the availability, scope and use of trade-related intellectual property rights;
(b) effective and appropriate means for the enforcement of such trade related intellectual property rights;
(c) the provision of effective and expeditious procedures for the multilateral prevention and settlement of disputes between governments; and,
(d) transitional arrangements aiming at the fullest participation in the results of the negotiations (Preamble).

The Agreement covers eight areas: copyrights and related rights, trademarks, geographical indications, industrial designs, patents and layout-designs of integrated circuits, protection of undisclosed information and control of anti-competitive practices in contractual licences. Part II of the

Agreement elaborates standards for the rights that need to be granted in these areas: the subject matter, the scope of protection, permissible exceptions as well as the duration of protection. When compared to earlier international intellectual property agreements, a major achievement of the TRIPS Agreement is the clear-cut prescription of minimum standards and scope of intellectual property protection (Lesser, 1995; Straus, 1996).

The provisions of the TRIPS Agreement have to be complied with by all WTO member countries, subject to certain exceptions. Articles 65 and 66 provide for different deadlines for implementation of the provisions of the TRIPS Agreement for developing countries and least developed countries respectively. Developing countries were allowed five years time to implement the TRIPS provisions in domestic environments, whereas the least developed countries have been given ten years to implement the same. In addition, there are some more exceptions in the Agreement that apply to pharmaceutical patents only. Article 65(4) of the TRIPS Agreement provides that in case countries did not provide for product patents at the date when the TRIPS Agreement came into force, they have the option of delaying product patent protection for pharmaceutical patents until 1 January 2005.[11] More recently, the Doha Declaration on TRIPS and Public Health (WT/MIN(01)/DEC/W/2) proposed an extension to the least developed countries to implement the health and pharmaceutical patent related provisions of the TRIPS Agreement until the year 2016. This extension has been accepted by the TRIPS Council in its June 2003 meeting.

The TRIPS Agreement and its impact on patents

There are several provisions in the TRIPS Agreement that strengthen the minimum standards of patent protection amongst WTO member countries, such as Articles 27(1), 27(3)(b), 28, 30, and 31(a) to (f) (Correa, 2001). The most important of these is Article 27(1) that makes it imperative upon Member States to make patent protection available "for any inventions, whether products or processes, in all fields of technology", and that "patent rights [be] enjoyable without discrimination as to the place of invention, the field of technology and whether products are imported or locally produced". Article 27(1) also specifies the three main conditions for an invention to qualify for a patent – novelty, inventive step (inventive progress) and the capability of being applied commercially.

According to the information available with the WTO Secretariat (Watal, 2001: 8), there were less than twenty developing and least developed countries that did not provide for product patents on pharmaceuticals and almost all of them provided for process patents when the TRIPS

Agreement came to force. Yet, the provisions of the Agreement on patents have had unequivocal impact on developing countries due to several reasons. In conjunction with one another, the provisions on patent protection in the TRIPS Agreement have the effect of eliminating several flexibilities that developing countries previously had on several aspects of patent protection. For example, prior to TRIPS, countries that provided only for process protection or for both product and process patents on pharmaceuticals had other ways of creating a balance between patent scope and public needs, such as the granting of compulsory licences when patents were not locally "worked". For example, in India, the government reserved the right to grant compulsory licences after three years (Reddy, 2003). The option to grant compulsory licences to "work" a patent locally is not possible now due to the procedural restrictions on licensing contained in Article 31 of the TRIPS Agreement. The Brazilian AIDS case could have probably shed light on this issue, had it not been withdrawn.[12]

As a result of stronger patent protection, unauthorized production or sale of patented drugs is illegal in all WTO member countries, unless they are produced after the expiry of the patent (Watal, 2001: 2).[13] Lastly, the Agreement also makes it obligatory for countries to grant intellectual property protection on life forms, a matter that was more of a discretionary nature before the Agreement.[14]

Introduction of intellectual property to newer areas

Article 27(3)(b) on *Patents on Life Forms* reads as follows:

> Members may also exclude from patentability:
> Plants and animals other than microorganisms, and essential biological processes for the production of plants or animals other than non-biological or microbiological processes. However, Members shall provide for the protection of plant varieties either by patents or by an effective sui generis system or by any combination thereof. The provisions of this sub-paragraph shall be reviewed four years after the date of entry into force of the WTO Agreement.

Article 27(3)(b) identifies five sorts of subject matter: namely, plants, animals, micro-organisms, non-biological and microbiological processes for the production of plants and animals, and plant varieties. It provides for different standards of protection for each respectively.

Article 27(3)(b) makes it obligatory for member countries to grant patents on micro-organisms and non-biological and microbiological processes for the protection of plants and animals, at the end of applicable transition periods.[15] Such patents on microorganisms and non-biological and microbiological processes for the protection of plants and animals can be claimed only when the three main criteria for patentability – novelty, inventive step, and industrial applicability – are satisfied.[16] But the Article gives countries the choice to deny patent protection to plants, animals, and plant varieties (Tansey, 1999: 8). However, Article 27(3)(b) requires that when countries deny patents on plant varieties, they provide an "effective" *sui generis* system for plant variety protection.[17]

The protection of microorganisms is the least disputed of the five subject matters identified here.[18] Protection of non-biological and microbiological processes for the production of plants and animals has been the cause of heated debate on cloning and embryonic technologies – a very critical issue but not the focus of this work. The impact of Article 27(3)(b) on plant varieties and *sui generis* systems has raised large questions about the availability of genetic resources for food and agriculture. The issue of the protection of traditional knowledge has also gained considerable prominence within the purview of this provision.

The link between intellectual property rights, biotechnology and biodiversity

Bioprospecting for drug research is one of the prime activities from which benefits can be shared in return for access to genetic resources and traditional knowledge (Straus, 1998: 115). It is primarily in this context that the intellectual property rights as contained in Article 27(3)(b) of the TRIPS Agreement interact with the provisions of the CBD on access, traditional knowledge and benefit-sharing in two major ways.

The first set of interactions is between conventional intellectual property rights and traditional knowledge. This raises two questions that are intricately interlinked: first, whether intellectual property rights present a satisfactory mechanism of protecting traditional knowledge and, second, whether intellectual property rights on biotechnological inventions aid the appropriation of traditional knowledge. Source countries, communities, and NGOs have been concerned for long that intellectual property rights as under the TRIPS Agreement encourage firms/private parties to claim patents on either genetic material per se, or traditional knowledge uses of genetic material without any obvious intellectual contributions. It has been felt that to the extent that they encourage such activities, such rights lead to misappropriation of traditional knowledge, like patents on traditional knowledge uses of turmeric, ayahuasca and neem. Although all this points out the need to protect traditional medicinal knowledge, should this protection mechanism be intellectual property and, if so, which kind of intellectual property?

A second set of interactions relate to the various stipulations contained in Article 16 of the CBD, which calls for the transfer of technology, including biotechnology, that may be essential to achieve the objectives of the CBD (Art. 16(1)). But it is unclear as to whether transfer of biotechnology is to be a form of benefit-sharing in bioprospecting contracts and, if so, under what conditions. Another part of this provision, Article 16(5), addresses the impact, both positive and negative, that intellectual property rights can have on the conservation and sustainable use of genetic resources. Article 16(5) consists of two parts – the first part invites parties to cooperate in ensuring that intellectual property rights are supportive of the objectives of the Convention and the second part sets out an obligation for parties to ensure that intellectual property rights do not run counter to the objectives of the Convention.

The first part of Article 16(5), which invites parties to cooperate in ensuring that intellectual property rights are supportive of the objectives of the Convention, calls for an assessment of the positive incentives that intellectual property rights could provide for the conservation of genetic resources. When seen in conjunction with Article 11 of the CBD, the potential medicinal or other values of genetic resources and the possibility of capitalizing on them through intellectual property rights can, in fact, act as a conservation incentive for certain groups of users.[19]

In the context of bioprospecting, the second part of Article 16(5), which calls for ensuring that IPRs do not run counter to the objectives of the CBD, translates into an examination of the link between drug research and unsustainable use of genetic resources. Source and user countries have been at loggerheads with each other vis-à-vis the harmful effects of intellectual property rights on sustainable use and conservation of biodiversity. Due to lack of process-oriented information in the international negotiation and law-making processes, actors have varied perceptions of whether medicinal products produced using biotechnological techniques create externalities in biodiversity use and, if so, to what extent IPRS on such products are to be held responsible.

Bioprospecting is the biggest victim of lack of clarity on these interactions because, without first settling these issues, the roles and responsibilities of users and providers cannot be set out in terms of national bioprospecting frameworks.

The TRIPS versus CBD debate and its implications for bioprospecting

It is now settled that there is no direct legal conflict between the provisions of the CBD and the TRIPS Agreement. Even if such friction did exist, the overriding nature of the TRIPS Agreement would hardly have provided a way out for developing countries to coin restrictions on use of traditional knowledge or access that bypass the stipulations of the TRIPS Agreement (e.g. Straus, 1998). As CEAS et al. (2000: 55) note, since TRIPS is the latter Agreement and the more detailed one for intellectual property protection, Article 30 of the Vienna Convention on the Law of Treaties would dictate that TRIPS prevail, since that is the Treaty that came later and focuses on intellectual property rights.

The lack of clarity with regard to the terms of the CBD, coupled with the diverse political ambitions of the biodiversity-rich source countries and biotechnologically competent user countries has paved the way for a tremendous concoction of issues related to sustainable use and conservation of biological diversity, the impact of intellectual property rights on these goals, and the issue of granting *sui generis* rights to traditional knowledge. Because of the long drawn-out controversy that it has been sparked off between countries on access and benefit-sharing, the Convention has come to be criticized largely for its ambiguity. As Miller (1997: 170) notes, "One of CBD's problems lies in the way in which its worthwhile goals are to be implemented. They are vague, ambiguous or impotent." As a result, the very ambitious attempt to integrate "previously distinct policy goals" in the CBD (Bragdon and Downes, 1998: 6) is on the verge of a failure. Key provisions such as Article 8(j), Article 15, and Article 16 are being used out of context and in isolation from the rest of the Convention. Many of the reasons for this lie in the politics that divide genetic-resource-rich source (yet developing) countries and technology-rich user (but developed) countries.

The politics of the TRIPS Agreement and the CBD

Even prior to the developments in biotechnology and the hype over the potential economic value of genetic resources, there was considerable friction in international relations on the terms for governing the use and exchange of plant genetic resources for agriculture.[20] This polarized the debate on ownership of genetic resources and the terms under which they should be exchanged for biotechnology-related use from the start (Stenson and Gray, 1999: 11). Developing countries were aggrieved that their genetic resources which formed the basis of many biotechnological inventions were being sold back to them at monopoly prices as a result of intellectual property protection, and saw Article 15 of the CBD as a means to assert control on their use and exchange.[21] Although the acceptance of Article 15 into the Convention draft was based on environmental reasons, it had strong economic undertones for both developing and developed countries (Mann, 1996). For many developing countries, it symbolized control over their "own" resources so that they could have a better say on the terms of their exchange. At the same time, developed countries also wanted genetic resources preserved from the point of view of their unrealized economic potential in the fields of medicine and other sciences. As Mann (ibid.: 26) notes: "The underpinnings of the main bargain between the North and the South in the biodiversity negotiations is often seen as the acceptance by the South of the obligation to conserve biodiversity and by the North of the obligation to share in the costs and benefits of this conservation. Central to this bargain are the provisions on the equitable sharing of benefits and rewards."

The way the TRIPS Agreement was negotiated did not help bridge the differences. From its inception, the Agreement has been tainted by limpid differences amongst countries about whether and to what extent intellectual property protection should be made part of global trade. The politics of the TRIPS Agreement stand proof to the fact that rapid developments in biotechnology was one key reason why developed countries were more interested in patent protection.[22] The Intellectual Property Committee that was formed to build consensus on the issue of incorporating intellectual property in the GATT was comprised of large industry interests, especially software and biotechnology.[23]

These divisions during TRIPS negotiations only served to reinforce the stand that the CBD provisions on access, traditional knowledge, and benefit-sharing are the primary tools through which developing countries can leverage benefits from bioprospecting and biotechnology-based drug research. For example, the question of protecting traditional knowledge had not been on the priority list of policy makers in most source countries until the advent of the CBD and the TRIPS Agreement and biotechnology-related developments that led to a sudden realization of the economic potential it presented.[24] But since the mid-1990s, the *sui generis* option under Article 27(3)(b) of the Agreement has been seen as a way to ensuring a property rights allocation on traditional knowledge that is more equitable from the perspective of developing countries. Subsequently, although Article 27(3)(b) allows member countries to provide for "effective *sui generis* system for plant variety protection", literature reveals two separate components of the entire *sui generis* option debate.

The first relates to the protection of plant varieties, which has been problematic in its own way for developing countries.[25] The second line of literature on *sui generis* systems under Article 27(3)(b) deals with the ways and means to protect traditional knowledge of local and indigenous communities in developing countries. This also includes traditional medicinal knowledge that is of relevance to drug research.

Defensive or positive protection of traditional knowledge?

Using examples of patents that were granted by Patent Offices in developed countries on several traditional knowledge uses such as turmeric and neem, proponents of "biopiracy" have time and again called for stronger access and traditional knowledge regimes to prevent Western firms from "stealing" traditional knowledge.

Yet it is hard to treat these cases alike. The turmeric case relates to knowledge that was already in the public domain within a specific region (in this case, India). The patent on turmeric was based on its use as known in India as a wound-healing agent, and therefore, was not novel. On the other hand, although the ayahuasca case was also revoked on grounds of novelty, it was nevertheless knowledge very closely held by indigenous communities in the Amazonian basin (as opposed to the use of turmeric that was common knowledge in India), and was embedded in a clear cultural context (Correa, 2001: 18). Therefore, although one could talk of both kinds of knowledge as "traditional knowledge" in a generic sense, the term "traditional knowledge" encompasses varying degrees of complexity, and varying degrees of contribution to modern medicinal systems. Proponents of a regime for traditional knowledge who do not make a distinction between the different kinds of knowledge (for example, those that are traceable to particular communities and those that are not; knowledge that is generally in the public domain versus knowledge that is not),[26] tend to simplify issues of enforcement and benefit-sharing, thus making sharing of benefits for such knowledge sound like mere rent-seeking.

Furthermore, such cases of "bad patents", as they are called, have to do with the construction of "prior art" that is used to examine novelty in certain jurisdictions rather than intellectual property rights as granted under the TRIPS Agreement.[27]

Novelty, inventive step, and industrial application are the three main requirements for patentability in patent jurisprudence. Article 27 (1) of the TRIPS Agreement also recognizes these three criteria for grant of patents. The novelty requirement is self-explanatory – an "invention" has to be new. The inventive step requirement implies that there be an evident intellectual contribution to the subject matter that warrants a patent. The requirement of industrial application implies that the invention on which the patent is sought is useful and applicable. To examine novelty, patent examiners have to usually ensure that the knowledge which is the basis of the invention is not in the public domain in any way. This requirement is known as "prior art".[28] Patents that are granted under some jurisdictions without the consideration of traditional knowledge as "prior art" have more to do with the respective national requirements on patent law than with the TRIPS Agreement itself, since Article 27(1) does not define novelty, inventive step, and industrial application.

To deal with these problems of construction of "prior art" and other interactions between traditional knowledge and intellectual property rights, two different protection models – positive protection and defensive protection – have been proposed.

Defensive protection refers to steps that are taken by national governments or regulatory agencies to prevent the misappropriation of traditional knowledge by unauthorized parties. Two modes of defensive protection have been suggested – a certification system and a system of improving prior art searches (Dutfield, 2004). The certification system has been fiercely debated in several forums, including the Conference of the Parties to the CBD (COP 7 and the ABS Working Group) and the Intergovernmental Committee on Genetic Resources, Traditional Knowledge, and Folklore of the World Intellectual Property Organization (WIPO).[29] A system of certification would mean that the national access authority (or an equivalent agency) issues a certificate of origin at the time of the access contract. In case the contract also requires the use of traditional knowledge, this may be accompanied by a certificate of prior informed consent from the holders of traditional knowledge.[30] But such a system needs consensus at the multilateral level, failing which it will not serve its purpose. For a system of certification to operate as an effective monitoring device, user countries would have to incorporate administrative requirements either: (a) in patent applications asking applicants to show proof of certificate of origin if the product in question contains genetic resources or traditional knowledge inputs; or (b) make civil or other penalties applicable to firms/individuals who do not adhere to the declaration of origin requirements. Source countries commonly argue at international negotiations that they can give "facilitatory" access only to countries that incorporate a PIC requirement in their patent laws making it easier for them to trace the flow of their genetic resources and traditional knowledge. Several source countries such as the African Group have even proposed that the requirement of declaration of origin be added as a conditionality for grant of patents, and that this should be made part of Article 29 of the TRIPS Agreement which sets out the rules on disclosure (Lettington and Nnadozie, 2003). User countries, on the other hand, are apprehensive about the interface of such traditional knowledge regimes with intellectual property rights on biotechnological inventions. For example, although the European Union is more open about such a system, it is not in favour of including such a requirement of origin as a patentability requirement (see for example, EU submissions to the WTO such as WT/GC/W/193). It has been felt that legal uncertainty regarding the protection of traditional knowledge (or the ambiguity in national laws on certain aspects of protection) may conflict with patentability of biotechnological inventions by firms. There has also been concern that including certification systems into patent laws would amount to adding a fourth requirement for the grant of a patent, thereby resulting in an alteration of minimum standards of patent protection.[31] Despite these differences, the Sixth Meeting of the Conferences of the Parties to the CBD has reiterated the importance of promoting measures that allow for disclosure of origin in patent applications (Decision VI/24). The Seventh Meeting of the Conference of the Parties has also been seen as the beginning of the consideration of a legally binding international ABS regime (Tobin and Johnston, 2004). Paragraph 19 of the Doha Declaration on TRIPS Agreement and Public Health (WT/MIN(01)/DEC/1) has instructed the Council for TRIPS to examine, inter alia, the relationship between the TRIPS Agreement and the CBD, the protection of traditional knowledge and folklore, and other relevant new developments raised by members pursuant to Article 71.1 of the TRIPS Agreement (see Correa, 2004).

Source countries have also voiced considerable support for the development of traditional knowledge databases as a means of improving "prior art" searches. This way, patent offices in countries that only search for prior art within specific geographical limits will be better equipped if such databases are made available to them.[32] But in doing so, source countries should be wary of the differences between registers of traditional knowledge and databases of traditional knowledge (WIPO, 2001).

Documenting traditional knowledge in such databases may convert a lot of oral traditional knowledge into a written form. These efforts can be used to create registers of traditional knowledge, a means of positive protection, whereby transaction-worthy information can then be converted into trade secrets or know-how licences (UNU-IAS, 2003: 11–12).

Positive protection of traditional knowledge refers to action taken by national governments in source countries to actively encourage the protection of traditional knowledge by recognizing rights over such resources. Whereas developing countries have showed willingness to explore intellectual property protection as an option for the protection of traditional knowledge and several national laws have already been enacted for this purpose, there have been considerable differences of opinion as to whether an effort should be made to develop an international *sui generis* system for the protection of traditional knowledge. The African Group has long called for bodies like the Inter Governmental Committee of the WIPO (hereafter, the IGC) to consider the establishment of an international *sui generis* system for traditional knowledge, which is of a mandatory nature.[33] Countries like India and the megadiverse countries have made similar submissions to the WTO and the

TRIPS Council.[34] At the same time, countries such as the USA have opposed the development of international rules for the protection of traditional knowledge at various forums including the WTO and WIPO (e.g. Correa, 2001: 21).

Article 16 of the CBD and its interactions with IPRs

The other interactions in the IPRs–biotechnology-biodiversity nexus are similarly strewn with controversy. Although technology transfer can theoretically include technologies that are meant for research and those that are primarily meant for conservation purposes, such as those that help monitor and classify biodiversity, it is unclear whether all of this would fall under the purview of the provision. Since the CBD has not specified these terms, considerable disagreement exists on whether it is only technology for sustainable use and conservation that source countries can demand in the form of benefit-sharing and, if so, under which circumstances. Whereas the developing countries feel that they ought to be able to choose and specify technology transfer and capacity building in exchange for their genetic resources, it has been argued that, based on Article 16(5), developed countries only have the obligation to provide or facilitate access of and transfer to technologies *related to conservation and sustainable use* (Straus, 1998: 106; emphasis added). Positions taken by most developing countries in ongoing international negotiations in various forums reveal a preference for flexibility in choosing the kind of technologies that they would like to have access to in return for genetic resources.[35]

Consensus on whether technology transfer can be demanded in return for access to genetic resources and traditional knowledge is being further compounded by two big realities of the drug R&D process. Much of the medicinal research being conducted today is for diseases faced by people in industrialized countries, such as Alzheimer's or Parkinson's. There is very little research on diseases such as malaria, tuberculosis and HIV/AIDS, which take the lives of millions of people every year in the developing world.[36] Owing to stronger patent protection contained in the TRIPS Agreement, there are concerns on drug accessibility and affordability in developing countries. The impact of patent protection on drug prices in developing countries is a complex issue and evidence available presently shows that these impacts cannot be generalized across countries.[37] However, recent experiences of developing countries in securing access to cheaper AIDS drugs for affected people go a long way not only to show their dependency on large pharmaceutical firms when patent protected medicines are urgently needed in their countries, but also demonstrate the complexities involved in making such medicines available at affordable prices. Both these factors make technology transfer a very imminent distributional issue in drug research that cannot be sidelined so easily.

The question of whether IPRs on biotechnological inventions run counter to the objectives of the CBD is also not settled. Source countries argue that in cases where medicinal uses of plants have been discovered, intellectual property rights lead to undue exploitation of such species and, therefore, users should be made to pay for biodiversity conservation. User countries tend to feel that such a demand for payment is acceptable but has to be linked to the harm caused (potential or actual). Owing to such different perspectives, the Conference of the Parties to the CBD has called for detailed assessments of the inter-relationships between intellectual property rights and the CBD goals in several instances.[38]

Resolving the interactions and setting the roles of users and providers

Setting political differences aside, resolving the legal controversies between the CBD and the TRIPS Agreement necessitates two big changes to existing analytical approaches. First, the role of the users and providers in access and traditional knowledge contracts can be conclusively decided only when the CBD is considered in its entirety. Otherwise, the fixing of "fair and equitable" benefit-sharing, deciding the roles of users and providers, and setting limitations on drug R&D on grounds of unsustainable use and conservation will at best be incomplete and imprecise.

Equally importantly, in order to enact optimal legal frameworks for bioprospecting, one needs to examine the details of the structure of the market for drug R&D based on genetic resources, the contractual problems faced by parties, the potential stages of R&D that may impact the sustainable use and conservation of genetic resources positively and negatively. Taking the market realities into account is also the best way to create a balance between the different stakeholder interests in the R&D process.

The right to access: The limitations

Within the cross-sectoral approach that the CBD envisages for biodiversity conservation and management, the right to access has a large role to play at the national levels. Upon a reading of the CBD, it is clear that no other authority has been created by the Convention to lead the cross-sectoral approach at national levels.

This role, although not so clearly laid out by the CBD, becomes evident when the different Articles are read in conjunction with one another. The three aims of the CBD – conservation, the sustainable use, and the fair and equitable sharing of benefits arising out of the commercialization of biological resources – are all to be achieved through the access mechanism. This link is evident from Article 1 which states, "including by appropriate access to genetic resources". Article 1 also recognizes the critical role that access laws play in ensuring that benefits are shared with communities within the country wherever their resources are used for commercialization purposes. This is contained in the next line of Article 1, which talks of "taking into account all rights over those resources". Therefore, the scope and functions of the right to access have to be decided based on the larger scheme of the CBD.[39] Other aspects of what the access authority should do and how it fits into the CBD scheme are contained in several other Articles of the Convention, the important ones of which are dealt with here. National access regulation, therefore, has to take shape within this scheme.

National biodiversity strategy, determination of focal points, and access laws

Subsequent to the CBD, three major tools have been identified as the key to achieve its objectives at national levels – a national biodiversity strategy, country study, and action plan (Miller and Lanou, 1995: 11). According to Article 6 of the CBD, before any access law is set up, a national biodiversity strategy has to be designed.[40] The devising of a national biodiversity strategy is seen as imperative in one form or another, as the Global Biodiversity Strategy has pointed out, to manage biodiversity as a whole (see again, WRI et al., 1992; see also McNeely and Vorhies, 2000: 9). The mandate of Article 6(a) – of devising a national biodiversity strategy that should reflect the measures in the Convention – translates into some main enquiries that national governments have to make, both at the normative and at the positive levels.

At a positive level, the national biodiversity strategy has to identify clearly the kind of biological diversity in the country in question, the pressures faced by reason of various user groups exploiting these resources simultaneously, and the strategies that could promote their sustainable use and prevent their extinction. On the normative side, the strategy has to identify the gaps between reality and the aspirations contained in the objectives, the issues and opportunities that the fulfilment of the objectives raises, the environmental impacts of the options available to attain the objectives and the kind of human, institutional, infrastructural, and financial capacities that are required to set the strategy in motion at the national level (Miller and Lanou, 1995: 11).

A country study, on the other hand, is "an assessment designed to gather information on the status and trends of the countries' species, genetic materials, and habitats and landscapes; on the status of current conservation and use mechanisms; and on the monetary and non-monetary costs and benefits involved" (Miller and Lanou, 1995: 12). Such a country study is to be a "first overview of the opportunities and problems of protecting and mobilizing a country's biotic wealth" (ibid.: 11). Article 7, sub-paragraph (a) of the Convention envisages precisely this.[41]

This information is important for the national biodiversity strategy because the information gathered through the country study is to be analysed in the light of the main pressures faced by these resources – and guides the potential goals and priorities identified by the national strategy.[42]

Based on all this, a national action plan has to be devised which sets out the national focal points in biodiversity conservation, as well as the steps required to implement the strategy and the institutions to coordinate the plan's goals at all levels – national, regional, and local (Miller and Lanou, 1995: 11). Articles 8 to 16 of the CBD provide guidelines to countries on these aspects. Sectoral plans have to be based on a framework that (a) identifies the basis of action (in accordance with Article 8); (b) identifies the process and activities; (c) identifies the components of biological diversity; (d) identifies the value of biodiversity; (e) identifies political, legal, economic, and administrative instruments (Miller and Lanou, 1995: 18–19).

A holistic reading of the CBD leaves little doubt that the access law itself should be wholly based on the priorities identified in the national biodiversity strategy and local priorities for sustainable use and conservation identified therein. Regulation of access can then prove to be a useful tool to generate human, institutional, infrastructural, or financial capacities that may be needed to implement the strategy over time.

Bioprospecting, access regulation, and benefit-sharing: The parameters

Even from a bioprospecting perspective, the resolution of the national priorities in biological resource use and conservation (of which genetic resources form a sub-set) should precede access laws because only then can fundamental issues be decided upon, such as the factors that should determine "appropriate" compensation, the biodiversity-related needs of the provider country that benefit-sharing should cater for, and the roles and responsibilities of users and providers.

Specifically, there are several aspects of bioprospecting frameworks that policy makers may be able to get better grips on, if access laws are designed pursuant to national biodiversity strategies and information gathered in the country studies. The documentation and information gathering carried out as part of the country study and the national biodiversity strategy are crucial to link the revenue generation aspects to conservation needs under the CBD. These can help in the creation of national inventories of medicinal plants, which, as Akerele notes, are essential if sound programmes for their rational use and exploitation are to be developed (Akerele, 1990). Such inventories need to describe the geographical and climate distribution of medicinal plants, enumerate their sources (collection from the wild, *in situ, ex situ* in botanical gardens, commercial plantations), and indicate their relative abundance or scarcity. For each plant there would be an account of its utilization (e.g. folk medicine, traditional healers, pharmaceutical and food industries) and its place in commerce (e.g. local use, international trade, export). There would also be a discussion of its constituents, pharmacological properties, and therapeutic indications (Akerele, 1990: 12).

The development of techniques aimed at a better valuation of genetic resources is also the key to incorporating the total costs of biological resource use in the local and national accounting. This is essential if cost-benefit analyses of preserving the genetic resources or conversion into its alternate uses should include the true social value of the resources (WRI et al., 1992: 22). Valuation techniques should also take into account the abundancy status of the biological resources and their ecological resilience (to recover from exploitation) to price them in national and international transactions. This would ensure that price-setting mechanisms in access procedures are more transparent. It is based on these parameters that the access authority should exercise its legal option on PIC when users, both nationals and foreigners, wish to use the resources within its territory.[43]

Information gathered in the country study and the inventories of medicinal plants can also help set the responsibilities of users in the sustainable use of genetic resources whose potential medicinal uses have been found (see Chapter 6 for a detailed discussion).

The "fairness and equity" consideration

Access authorities and local and indigenous communities are rights-holders according to the text of the Convention itself, thereby mandating that the national laws and institutions that are designed to implement the CBD take into account these rights and provide mechanisms to enforce them. The accurate prescription of roles means that access authorities recognize the communities as legitimate parties to bioprospecting contracts, wherever applicable (that is, whenever firms/private entities would like to transact for traditional knowledge) and enable their participation in settling upon MAT, as Article 15 requires. Access institutions have the big onus of representing communities that possess traditional knowledge in bioprospecting contracts effectively. MAT would require that these communities have sufficient autonomy and control on their knowledge to exercise their consent.

But there are also the interests of other stakeholders (all user groups) that the national access laws have to take into account, and this aspect is not excluded by the Convention. Article 11 recognizes the issue of granting incentives and, as already mentioned, this extends to *all* stakeholders, whether national or foreign. Only this can promote economic investments into biological diversity in a sustainable way. Access laws have to prescribe roles keeping in mind the prominent role that revenues generated through sustainable use can have in programmes for biodiversity conservation.

Thus, when the Convention is taken in its entirety, the "fairness and equity" mandate has two dimensions to it. On the one hand, it demands that just as the demanders of access be obliged to reveal information and pay compensation for the use of the resource, the access and traditional knowledge laws too must employ transparent and user-friendly mechanisms of price-fixing for genetic resources. The "fairness and equity" requirement cannot be translated into reality when either the access regulations that incorporate traditional knowledge into itself or independent traditional knowledge regimes do not provide for the same transparency and accountability criteria as between the national governments and local and indigenous communities or other parties, such as private landowners, who may be the actual owners of the land to which access is sought. Absence of this sort of accountability and clarity between access and traditional knowledge regimes may also lead to creation of false expectations amongst parties about their share, nature of the resource, or their bargaining thresholds, the avoidance of which should be the main aim of policy making. This, too, hinges largely upon well-defined property rights.

The right to traditional knowledge: The limiting factors

Traditional knowledge, as defined under Article 8(j), lays down the conceptual elements of what the right should be and what national governments should endeavour to protect. But Article 8(j) in itself cannot be the definition for a "right" to traditional knowledge or its institutional mechanism for rights' enforcement.[44] Article 8(j) comes with two caveats. First, it makes national governments responsible for the implementation of its content,[45] which implies that countries have to devise their own definitional limitations to the concept according to: (a) the nature of local and indigenous communities, and their knowledge holdings; (b) general organizational limitations on implementing a community intellectual property right; and, finally, (c) specific organizational limitations in implementing the right that may be peculiar to that country itself. Second, the Article is titled "*In situ* conservation", implying that any law to protect this knowledge should clearly be able to link it to conservation.

Clarifying the interface between Article 27(3)(b) of the TRIPS Agreement and traditional knowledge

Although a detailed reading of Article 27(3) and all its provisions leaves one desiring more clarity in the definitions,[46] some aspects do stand out. Article 27(3)(b) permits a *sui generis* option for plant variety protection. The only rider that the Article contains for the *sui generis* option is that it be "effective".[47] Given that the TRIPS Agreement only prescribes minimum standards of intellectual property protection, member states seem to have the option to determine what an effective *sui generis* system for plant variety protection should look like, subject to the general restrictions of the Agreement.

Article 8 on "Principles" also contains the general restrictions placed on countries when devising additional provisions. The general restrictions of the TRIPS Agreement as contained in Article 8 are that the *sui generis* right:

(a) should not intervene with the minimum standards of intellectual property rights as set out under the Agreement; and,
(b) should not result in any complications that may result in a lesser compliance with the terms of the Agreement, as such.

As long as these two restrictions are conformed to, the *sui generis* right can be one that suits local conditions best.

Article 27 itself contains two other exceptions that countries can rely on when devising national regimes for the protection of life forms – Article 27(2) and Article 27(3)(a). As a result of Article 27(2),[48] an exception on the grounds of animal, human, plant life, and the environment is permitted. But what exactly is meant by "*ordre public*", and "morality" is not defined once again.[49] It has been felt that the implementation of Article 27(2) at the national level mandates the resolution of the meaning of the term "*ordre public*" – which could either be interpreted as being mainly linked to security reasons (as under the Guidelines for Examination of the European Patent Office) or as a general public order or public interest.[50] A limitation to the exception under Article 27(2), though, is that Members are not allowed to refuse patent to a commercial invention merely because it is prohibited under national law. According to Article 27(3)(a), the diagnostic, therapeutic, and surgical methods for the treatment of humans or animals can be excluded.[51]

But even despite all these ambiguities, it is still difficult to see how exactly Article 27(3) and its provisions are to hinder countries directly from enacting their own traditional knowledge laws provided that it does not minimize the impact of the Agreement in any way. In this regard, Article 8(1) is a noteworthy provision of the Agreement, at least to the extent of providing some flexibility to countries in implementing the provisions of the TRIPS Agreement. It is generally accepted by now that the "Public Interest" principle embedded in Article 8(1) allows member states, when formulating or amending their national laws and regulations on intellectual property, to adopt measures necessary to protect public health and nutrition; and to promote the public interest in sectors of vital importance to their socio-economic and technological development. Naturally, the proviso to this freedom is that such measures should be consistent with the provisions of the Agreement as such, as already discussed.

Balancing the various interests: The relevance of a well-defined right

Regardless of whether traditional knowledge is considered independently, or as a part of the access and benefit-sharing framework, defining the right to traditional knowledge in a way that its implementation is feasible seems to be the most important policy task.

The nature and composition of local and indigenous communities and their knowledge holdings is a concept more heterogeneous than one generally imagines. Traditional knowledge should be further broken up into categories that recognize the plausible contribution of the knowledge to modern research, and for each one of these cases, right holders and beneficiaries should be identified and compensation mechanisms should be devised. The definition of the resource should be able to clarify the nature of the resource, the scope of protection, and the beneficiaries in question.

Apart from this, the rights' definition has to take into account general organizational limitations such as principal-agent problems between the communities and the state agencies, such as the access authority that represents them, ways to organize this relation so as so ensure maximum transparency, and community participation, etc. Specific organizational limitations of countries could also come into play depending on sociocultural differences that may be dominant enough to affect the implementation process; for example, in the Indian context, caste and creed differences within communities may be another factor that legal policy makers have to really take into account.

Furthermore, the emphasis that such a regime should place on conservation incentives should not be understated. Countries are responsible to ensure that such legislations do end up granting these groups sufficient incentives to conserve biological diversity in an optimal way.

Notes

1. The Global Biodiversity Strategy was a pre-emptive initiative to the CBD that sought to induce a fundamental paradigm shift on conservation and sustainable use strategies for biodiversity at the international, national, and local levels.
2. This flexibility has been appreciated largely for its "result-oriented approach". See, among others, Burhenne-Guilmin (1994: 16).
3. The kinds of incentives that can be provided for biodiversity use and conservation is a topic that has been extensively discussed in economic literature. For a discussion of the kinds of incentives one can envisage for biodiversity and their implementation, see for example, OECD (1995: 38–39; 79–81).
4. The Convention 169 concerning Indigenous Peoples in Independent Countries was the first Convention to deal with this issue. For an extensive analysis of the development of rights of indigenous peoples and the international legal instruments related to it, see Posey (1996).
5. A detailed assessment of binding and non-binding international agreements, customary practices, moral and ethical principles shows that a rich body of rights exist for indigenous people. It has been suggested that countries enact national regimes for indigenous people using this "bundle of rights" as the guideline. For an exhaustive review of rights of indigenous people in international law, see Posey (1996).
6. Article 15(5) reads as follows: "Access to genetic resources shall be subject to prior informed consent of the Contracting Party providing such resources, unless otherwise determined by that Party."
7. See Mann (1996: 26–30), who notes this after an extensive review of existing literature on PIC.
8. Article 15(4) reads as follows: "Access, where granted, shall be on mutually agreed terms and subject to the provisions of this Article".
9. Article 15(2) reads as follows: "Each Contracting Party shall endeavour to create conditions to facilitate access to genetic resources for environmentally sound uses by other Contracting Parties and not to impose restrictions that run counter to the objectives of this Convention."
10. The range of options that can be considered for benefit-sharing have been laid out by the Bonn Guidelines on Access and Benefit-Sharing (see Appendix II of the Bonn Guidelines on Access to Genetic Resources and Fair and Equitable Sharing of the Benefits Arising out of their Utilization, Decision VI/24, UNEP/CBD/COP/6).
11. The Indian generic industry is a clear example of how countries have used this exemption until now.
12. In the Brazilian case, the USA brought a complaint against Article 68 of Brazil's Industrial Property Law of 1996 in front of the WTO dispute settlement panel. Specifically in question was the legality of the provision that provided that a patent should be made subject to compulsory licence if it was not "worked" locally. The US claimed that such a provision contravened Article 27(1) of the TRIPS Agreement since it amounted to discrimination against patent holders who imported products into the Brazilian market as against those who produced it locally. Before the WTO Dispute Settlement Panel could decide on the issue, the USA withdrew the case based on a bilateral agreement with the Brazilian government. See Gehl Sampath (2004).
13. This restriction only applies to countries in which the said drug is patented; in countries where the drug is not patented, generics can be produced and/or sold, since patents only apply territorially.
14. Prior to the TRIPS Agreement, the protection of life forms was controversial, although allowed under some jurisdictions. The Union for Protection of Plant Varieties (UPOV),

1972 allowed for breeders' rights on plant varieties and several countries were part of the Union. In the USA, patents on microorganisms were allowed after *Diamond* vs. *Diehr* 450 US 175 (1981).

15. Articles 65 and 66 provide for different deadlines for implementation of the provisions of the TRIPS Agreement for developing countries and least developed countries respectively. Developing countries were allowed five years' time to implement the TRIPS provisions in domestic environments, whereas the least developed countries have been given ten years to implement the same. See also Note 12 for some other exceptions.
16. For an invention to qualify for a patent, there are three main requisites: it has to be novel, it should possess an inventive step (inventive progress), and it should be capable of being applied commercially. Article 27(1) of the TRIPS Agreement specifies these three requirements clearly, but does not define them. As a result, deciding what constitutes novelty, inventive step, and commercial application is a matter of national jurisdiction, although intellectual property jurisprudence contains some common notions. For example, generally, "prior art" is the criteria applied by patent offices to judge novelty – for an invention to be novel, it should not be available in the public domain.
17. "*Sui generis*" simply denotes a system of its own.
18. But since the Article does not define "microorganisms" amongst others, some feel that it is natural to conclude that this would also include extending the protection to similarly occurring natural substances like cell lines (see Straus, 1998: 1090). For an opposite view, see Correa (2000: 68).
19. According to Article 11, each contracting party shall, as far as possible and as appropriate, adopt economically and socially sound measures that act as incentives for the conservation and sustainable use of components of biological diversity.
20. The International Undertaking (IU) on Plant Genetic Resources, adopted by member countries under the FAO commission in 1983, was the main intergovernmental undertaking on plant genetic resources until the CBD. It was non-binding in nature and recognized that "plant genetic resources are a heritage of mankind to be preserved, and to be freely available for use, for the benefit of present and future generations". The IU was conceived in an atmosphere of severe disagreement between developed and developing countries on ownership of plant germplasm, plant breeders' rights, and rights over *ex situ* germplasm collections.
21. Authors like Shiva called biotechnology a means of establishing what they term "neo-imperialism", by making the developing countries once again completely dependent on the industrialized countries and by allowing the latter to exploit the former's resources without repatriation of benefits (see Shiva, 1997).
22. On this issue, Tansey (1999: 6) notes, "One reason for greater interest in patents is the rapid development of biotechnology, especially the OECD countries, and its application to agriculture". The changing nature of innovation in fields like software and biotechnology (characterized by huge capital investments and possibility of easy replication) was accompanied by the feeling on the part of the industrialized countries that trade gains that result from inadequate intellectual property protection should be curbed. Straus (1996) notes, "the anomaly of an almost lack of minimum rights particularly in the field of patents weighed against principles of territoriality and national treatment was not acceptable – to let certain countries to participate in both technological advances and globalization and the integration of markets and at the same time to appropriate inventions free of charge at their national borders, regardless of the use that could be made of them". For other discussions on country positions during TRIPS negotations, see Cottier (1991), CML Rev. 386; Pacón (1996), Braithwaite and Drahos (2000).
23. As Drahos (1995) notes, the Intellectual Property Committee that was formed to gener-

ate consensus for the US plan to include intellectual property into GATT in Europe, Canada, and Japan (the so-called QUAD Group) consisted of Bristol-Myers, Johnson and Johnson, Merck, Pfizer, Monsanto, General Motors, IBM, Hewlitt Packard, General Electric, Du Pont, FMC Corporation, Rockwell International, and Warner Brothers. See Drahos (1995: 6–18, at p. 12).

24. See Gopalakrishnan (2002) who discusses this in the case of India, where even now it seems that the priority of legislative attempts is more to attract the industry rather than to protect and empower indigenous and local communities.
25. An option that has been advocated in this regard is the 1991 version of the Convention of the International Union for the Protection of New Varieties of Plants (UPOV). There has been concern that the UPOV Convention may not be best suited to address the needs of farmers in developing countries due to various reasons. Farmers in developing countries rely on annual seed saving to a very large extent. The UPOV Agreement, in this regard, provides for an optional provision on "farmers" privilege'. According to Article 15 of UPOV Convention 1991, if Members wish to, they can permit farmers to replant specific varieties for propagating purposes within their own farm holdings, out of the products that they obtained from planting such a protected variety in the first place. However, the Agreement urges Member Countries that this privilege should be granted to farmers only "within reasonable limits and subject to safeguarding of the legitimate interests of the breeder", because if the privilege were interpreted in a very broad way, it might affect the incentive of breeders to invest in the creation of new varieties (see Article 15 again). Moreover, Article 15 of the UPOV Convention does not permit for seed exchange, a practice that is very common amongst small farmers in developing countries. There has also been concern that the definition of a "breeder" in UPOV 1991 Convention allows for the undue appropriation of the germplasm of farmers of the Southern countries because in the UPOV 1991 version, a "breeder" includes someone who "discovers the plant variety".
26. Information that is in the "public domain" is generally understood as information which is potentially available to every member of the public.
27. In the literature, the term "bad patents" is used to denote patents which are granted without conforming to the novelty requirement.
28. According to the WIPO (2001), the term "prior art" generally refers to the entire body of knowledge which is available to the public before the filing date or, if priority is claimed, before the priority date, of an application for certain industrial property titles, principally patents, utility models, and industrial designs.
29. For example, see the Proceedings of Sixth and Seventh Conferences of the Parties to the CBD, April 2002, at: www.biodiv.org. For a discussion of the aspects of certification being negotiated under the international ABS regime, see Tobin and Johnston (2004). The IGC has considered the certification of origin system at length. See for example, WIPO/GRTKF/IC/4/11 and WIPO/GRTKF/IC/5/11. WIPO/GRTKF/IC/5/6 also contains a list of various options that countries can use for the defensive protection of traditional knowledge too. For a discussion of these options, see Lettington and Nnadozie (2003: 17–18).
30. As Dutfield (2000: 77) notes, according to this proposal, administrative requirements for filing patent applications based on the use of genetic resources and/or traditional knowledge should require (a) a sworn statement as to the genetic resources and traditional knowledge, innovations, and practices of indigenous peoples and local communities, utilized directly or indirectly in the research of the subject matter of the patent application; as well as, (b) evidence of PIC from the country of origin and/or indigenous peoples and local communities, wherever applicable.
31. The European Council proposed Amendment 76/rev to the EU Directive 98/44 which

provided that a patent will be granted only when the details of the geographical origin of the material and proof of legal access in accordance with the rules of the place of origin are provided by the patent applicant. This was rejected by the Parliament since most Member Countries felt it ran contrary to Article 27(1) of the TRIPS Agreement (Lightbourne, 2003: 889). See also, Wolfrum et al. (2001: 19) for a similar argument.

32. See for example, that Sections 102(a) and (b) of the US Patent Act contain limitations on geographical scope, as a result of which only "prior art" that is written in nature and available within the USA is considered by the patent examiners in the USA. In contrast, according to Article 56 of the European Patent Treaty, the European Patent Office is open to consider all "prior art", whether oral or written.
33. See for example, WTO, IP/C/W/163, Submission by Kenya on behalf of the African Group that proposed including a footnote to Article 27(3)(b) to provide for community rights on traditional knowledge. See also Correa (2001); Lettington and Nnadozie (2003).
34. The coalition of the Like Minded Megadiverse Group of Countries composes of Bolivia, Brazil, China, Colombia, Costa Rica, Ecuador, India, Indonesia, Kenya, Malaysia, Mexico, Peru, South Africa, and Venezuela. See for example, submission WT/GC/W/251 of 13 July 1999 where the Dominican Republic, Egypt, El Salvador, Honduras, India, Indonesia, Malaysia, Nigeria, Pakistan, Sri Lanka, and Uganda proposed that Article 27(3)(b) should be amended in the light of the provisions of the CBD (and the IU, which is now the International Treaty on Plant Genetic Resources) in which the rights of local and indigenous communities on traditional knowledge is fully taken into account.
35. See again the negotiations on access and benefit-sharing as part of the Sixth Conference of the Parties to the CBD.
36. As *The Economist* (10 September 2001) notes, "Most of medical research is done in rich countries, for the benefit of the rich people. The fattest profits are to be made from tackling chronic conditions that affect a lot of westerners, such as heart diseases and cancer. The ills of the poor are neglected: of the 1,223 drugs introduced between 1975 and 1996, only 13 were aimed at tropical diseases. In 1998, the world spent US$70 billion on health research but only US$300 million of this was spent on developing an AIDs vaccine and a piffling $100 million was directed at malaria research."
37. There are several factors that determine whether stronger patent protection will in fact lead to price rises in developing countries, such as the structure of the local industry, presence of therapeutic substitutes, local demand elasticity, and the presence of price controls. For more, see Gehl Sampath (2003). Depending on the presence or absence of these factors, the pinch on drug availability and affordability will be felt differently in different developing and least developed countries.
38. See in this context, COP Decision III/7 which called for case studies on the impact of IPRs on the achievement of CBD objectives; see Decision IV/15 which calls for appreciation of the relationship between IPRs and the CBD; see Decision VI/24 which contains an entire section on the role of IPRs on the implementation of access and benefit-sharing provisions.
39. Many COP Decisions have reiterated the same. For example, Decision III/15 noted that the implementation of Article 15 is closely linked to other Articles of the CBD.
40. According to Article 6:

    Each Contracting Party shall, in accordance with its particular conditions and capabilities: (a) Develop national strategies, plans or programmes for the conservation and sustainable use of biological diversity or adapt for this purpose existing strategies, plans or programmes which shall reflect, inter alia, the measures set out in this Convention relevant to the Contracting Party concerned; and (b) Integrate, as far as possible and as appropriate, the conservation and sustainable use of biological diversity into relevant sectoral or cross-sectoral plans, programmes and policies.

41. Article 7(a) reads as follows: "Each Contracting Party shall, as far as possible and as appropriate, in particular for the purposes of Article 8 and Article 10, identify components of biological diversity important for its conservation and sustainable use having regard to the indicative list of categories set down in Annex 1."
42. This again is in accordance with Article 7(c) which reads as: "Each Contracting Party shall, as far as possible and as appropriate, in particular for the purposes of Article 8 and Article 10, identify processes and categories of activities which have or are likely to have significant adverse impacts on the conservation and sustainable use of biological diversity, and monitor their effects through sampling and other techniques."
43. See Akerele, who notes that even prior to the Convention, the inter-regional consultation sponsored by WHO and the IUCN and the WWF in 1988 stressed upon the importance of including measures for conservation in countrys' investigation, utilization and exploitation of medicinal plants Akerele (1990: 13).
44. See Burhenne-Guilmin and Glowka (1994) in this regard who note that the framework of the Convention is result-oriented, leaving a lot flexibility for Parties to decide how to put its mandate into action within their respective national situations (p. 16).
45. The Article starts with: "Subject to national legislation ...".
46. On the issue of multiple interpretations of the terms-plants, animals, microorganisms, essential, non-biological, microbiological, plant varieties, effective and *sui generis* system, see Tansey (1999: 23–25).
47. According to Article 27(3)(b), "Members shall provide for the protection of plant varieties either by patents or by an *effective sui generis* system or by any combination thereof" (emphasis added).
48. Article 27(2) reads "Members may exclude from patentability inventions, the prevention within their territory of the commercial exploitation of which is necessary to protect *ordre public* or morality, including to protect human, animal or plant life or health or to avoid serious prejudice to the environment, provided that such exclusion is not made merely because the exploitation is prohibited by law.
49. Dutfield notes in this regard that the wording of Article 27(2) is very similar to that of the European Patent Convention (Art. 53, in particular), and that even within Europe, the true meaning and content of the terms "ordre public" and "morality" is unresolved. See Dutfield (2000).
50. See the discussion in Correa (1998: 192).
51. According to Article 27(3)(a) of the TRIPS Agreement: "Members may also exclude from patentability: diagnostic, therapeutic and surgical methods for the treatment of humans or animals".

4

Transaction costs and their impact on the market for bioprospecting

I. INTRODUCTION

The efficiency of allocation in a perfect market requires a competitive contracting process with perfectly informed and rational utility-maximizing individuals, the absence of transaction costs, and the absence of externalities to third parties (Kreps, 1995). The kind of transactions in such a perfectly competitive market is what lawyers refer to as classical contracts, or what scholars of law and economics call complete contracts (Williamson, 1991: 271). In a classical contract, the identity of the parties is irrelevant because the nature of the agreement is overbearing, the formal written features of the agreement govern the more informal oral understanding when a contestable situation arises and the remedies for breach are narrowly prescribed (see Williamson, 1979, 1991).[1] Although fundamentally, in both law and economics, contracts perform the same task, i.e. facilitation of exchange, whereas lawyers consider this a mechanism for extracting binding commitments from people, for economists it is primarily a mechanism for alignment of incentives to ensure optimal performance (Milgrom and Roberts, 1996).

At the other end of the spectrum are what economists call incomplete contracts or contracts that would fall under the category of neoclassical contract law as per legal understanding (Williamson, 1991). These are essentially *long-term* contracts executed under conditions of *uncertainty* for which complete presentation of all risks, rights, and obligations are prohibitively costly, if not impossible (Williamson, 1979: 237). In such contracts, contractual problems arise due to (ibid.):

(a) failure to anticipate all future contingencies for which adaptations in the original bargaining thresholds are required at the time of formulating the agreement;
(b) impossibility to conceptualize the kind of adaptations that may be required to be made in case of specific contingencies, until these said contingencies actually materialize;
(c) difficulties in resolving legal disputes because verification of the claims of the contractual parties by outsiders is not as easy a task as classical contract law imagines.[2]

These constraints impose costs on the contractual parties in contrast to complete contracts. Such *transaction costs* are the costs associated with transfer of goods (or rights) from one individual or juridical person to another. Transaction being the most fundamental unit of exchange within economic theory, transaction costs refer to any or all costs that emerge within such "trade" that parties have, in order to exchange goods of any sort. They include, in addition to routine costs, other costs which are basic to establishing and maintaining a basic institutional framework that facilitates exchange, or "the costs of running the economic system" (Arrow, 1969: 48).[3]

Transaction cost economics has identified bounded rationality, uncertainty, opportunistic behaviour, asset specificity and frequency of the transaction to be the main characteristics of a transaction that give rise to transaction costs. Normally, transaction costs can be anything between negligible to extremely significant for a given transaction depending on whether upon the presence of these characteristics, the inability to foresee each and every contingency and to conceptualize adaptations for them are inconsequential, consequential, or highly consequential (Williamson, 1991: 272). In the case of incomplete contracts, such disturbances are at least consequential in nature, as a result of which the continuance of the contractual relation to create mutual gains is conditional on the adaptive capabilities of both parties to emerging situations. Incomplete contracts therefore require highly adjustable and adaptable contractual structures, as opposed to classical contracts that are complete and contractual structures that can be narrow and rigid (ibid.).

The main aim of this chapter is to analyse the structure of bioprospecting contracts and link it to the underlying economic problems the parties are faced with. If bioprospecting contracts are not ideal transactions/classical contracts concluded under perfectly competitive markets, the market conditions that govern the exchange of genetic resources and traditional knowledge are critical in determining which forms of contractual structures and policy interventions may be required. As Chapter 2 has already enumerated, there are several kinds of contractual structures in bioprospecting. The various modes of organizing the same transaction raises a fundamental question: what are the factors that motivate parties to choose one contractual structure over another one?

The market for bioprospecting cannot be described as a perfectly competitive market, but rather one that exhibits various imperfections. Bioprospecting for drug research, just like any other drug R&D activity, is characterized by uncertainty because none of the parties can actually predict whether a drug will be discovered at all, and, if so, under what circumstances. Given this, most of the contractual negotiations, especially those related to benefit-sharing, are conducted against the backdrop of uncertainty. If the contracting parties wish that benefit-sharing stipulations are commensurate with the results of the process, it is probable that the stipulations need to be revisited at each stage of the process as and when more information about research prospects become available. Bounded rationality in the context of bioprospecting implies that, since uncertainty prevails and the R&D process and its results are unpredictable, even if the parties behave rationally, their information about the future is imperfect. In order to make the contracts more complete, parties may have to gather more information about all future prospects and contingencies that could potentially materialize to jeopardize their contractual bargain, and this entails costs. Failing to organize transactions in a way so as to deal with uncertainty and bounded rationality also exposes the parties involved in bioprospecting contracts to opportunistic behaviour.

These market imperfections can be described under two categories. The first category consists of problems of opportunistic behaviour due to asymmetric information that can be divided into pre-contractual opportunism (misrepresentation of value and quality of both, samples of genetic resources, and goals of R&D programmes) and post contractual opportunism between the parties (principal-agent and monitoring relationships). The second category of *ex post* contractual imperfections relate to problems of asset specificity, which are caused by opportunistic behaviour or due to legal uncertainty and imperfect enforcement issues.

In the presence of these market imperfections, spot market transactions for genetic material or traditional knowledge that one sees in this market cannot be interpreted as optimal contracts in a perfect market, but rather as transactions that reflect the lost potential of exchange. Such contractual exchange is proof of low-level transactions lacking complexity mainly because high-level transactions with complex contracting that would actually be needed cannot emerge/be enforced in the market under present constraints posed by ill-defined or non-existent property rights and insufficient legal regulation (Hart, 1995; Hart and Moore, 1990).

Efficient allocation of goods (or more generally, property rights) under the constraint of transaction costs, the so-called second-best solutions, can be obtained by designing sophisticated contractual provisions rather than those based solely on price versus quantity parameters (also called linear or uniform pricing). Certain kinds of contractual provisions are optimal responses to transaction costs. Good examples are benefit-sharing agreements linking the price for goods/services to future market outcomes of final products, long-term contracts to share investments, specialized firms with established reputation that indulge in repeated contracting, joint ventures as a model of firm cooperation, and partial vertical integration.

To analyse these market imperfections in a systematic way so that appropriate property rights definitions and scope for legal regulation can be predicted, this chapter is structured as follows. Section II uses transaction cost economics as a morphological tool to describe the kinds of transaction costs and market imperfections that parties face in the market for bioprospecting. But since transaction cost economics has its own limitations in quantifying the effects of transaction costs and proposing appropriate solutions, the remaining sections in the chapter use economic theories of contracting and information economics to discuss potential solutions. Section III deals with what can broadly be called search and information costs, Section IV discusses bargaining and decision costs, and Section V focuses on monitoring and enforcement costs. Section VI deals with imperfect commitments and specific assets, and its relevance to bioprospecting collaborations.

II. TRANSACTION COST ECONOMICS AND ECONOMIC THEORIES OF CONTRACTS: AN APPLICATION TO BIOPROSPECTING CONTRACTS[4]

This section presents an overview of transactions cost economics and assesses whether these costs are present in bioprospecting contracts. The review of transaction cost economics in this section and economic theories of contracting presented in the rest of this chapter set the stage for applying the results to concrete property rights definitions and contractual solutions in the market for bioprospecting.

A closer look at the assumptions of transaction cost economics

According to transaction cost economics, when the behavioural factors that govern a transaction are opportunism and bounded rationality, and when the transactional factors are uncertainty, asset specificity, and frequency, parties face issues of transaction costs (see generally, Williamson, 1975, 1979, 1991, 2000).

Behavioural factor I: Opportunistic behaviour

The idea of opportunistic behaviour is central to transaction cost economics and is defined as "self-seeking with guile" (Williamson, 1975). Fundamental to the understanding of what constitutes "opportunistic behaviour" is the distinction between simple self-interest and opportunism. Simple self-interest would apply to any completely honest individuals, who would never misrepresent a fact or break their promise, but who seek to maximize their welfare within given constraints. In contrast, opportunistic behaviour demands that a person refuses to divulge important information or even give false information when they know that the other party has no means to verify it (Kreps, 1995: 745). Therefore, for any behaviour to be called "opportunistic", actors have to follow their own self-interest to the disadvantage of others and, in doing so, neglect the generally accepted rules of mutually beneficial exchange (ibid.).

Opportunistic behaviour is especially problematic in the case of public goods, trading in information, and complex contracts where each contingency is not *ex ante* provided for (Williamson, 1975, 1991).

Behavioural factor II: Bounded rationality

Neoclassical economics assumes the presence of rationality as pure logical axiomatic behaviour. According to neoclassical economics, individuals know everything that they ought to know and they process this information and form preferences according to certain set axiomatic rules – completeness, transitivity, and reflexiveness (see Arrow, in Mas-Colell et al., 1995: 6). Utility functions are derived from individual preferences ordered in this way. Neoclassical economics also assumes that a rational economic actor will behave in such a way so as to maximize his (expected) utility under given constraints.[5] This notion of rationality in neoclassical economics is not linked to the availability of information. Therefore, individuals can be rational even with incomplete or imperfect information – as is modelled in information economics generally.

Transaction cost economics assumes bounded rationality amongst individuals. As opposed to the notion of rationality in neoclassical economics, bounded rationality, a fundamental informational assumption of transaction cost economics, lacks a definition of formal rigour.

Bounded rationality simply means that individuals operate with limited information sets wherein either not all contingencies are known and/or cannot be written into a contract (Kreps, 1995: 744). This is in contrast with the assumptions of neoclassical economic decision theory, although individuals do not violate any of the rationality axioms based on the limited amount of information available to them. As Willamson notes, bounded rationality should not be confused with irrationality; it merely refers to a situation wherein human agents are "*intendedly* rational, but only *limitedly* so" (Simon, 1961, cited in Williamson, 1975: 126). In day-to-day jargon, this simply means that individuals can make errors in decision-making because they base their decisions on limited information or sometimes on assumptions that may be erroneous or may not materialize in reality.

In the world of transaction cost economics, the challenge is to organize transactions that economize on bounded rationality considerations while simultaneously safeguarding the transactions against opportunistic behaviour of others (Williamson, 1975: 127). Bounded rationality and opportunism themselves do not amount to much when the sort of transaction that is contemplated is an extremely simple one, such as a spot market transaction (Kreps, 1995). The following three aspects of a transaction are the ones which work hand in hand with bounded rationality and opportunism to cause transaction costs.

Transactional factor I: Uncertainty

Uncertainty is a transactional parameter that actually defines the information assumption of bounded rationality (Kreps, 1995: 749). Uncertainty denotes the situation where people have to make decisions without knowing for certain what the consequences of their decisions will be (see Katz and Rosen, 1998). In other words, they cannot know at the time of their decisions in how many states of the world the said event/result will materialize, or what contingencies may prevent the materialization of that desirable state. This makes it too costly to envision and incorporate the potential contingencies into a contract. The concept of uncertainty in transaction cost economics also includes the kind of uncertainty that is caused when one party to a contract has more information about the exchange than the other (Kreps, 1995: 749). As a result, implementing a contract under uncertainty requires that parties agree to renegotiate a contract contingent on any new state of the world, if and when unforeseen changes in the environment do occur. Such renegotiations become problematic if parties make relation-specific investments under the original contract, as will be discussed later on in this chapter.

Transactional factor II: Asset specificity

Assets are said to be specific in relation to a transaction if the asset is worth much more within the transaction than outside it. In such cases, asset specificity arises when parties cannot write complete contracts that specify the use of assets in every contingency during the course of the contract. Parties can be held up if they cannot switch partners at negligible costs after investing into the relationship (Hart, 1995: 25). The concept of quasi-appropriable value is central to understanding asset specificity and the problems associated with it. The quasi value of an asset is its value in its next best use (Klein et al., 1978: 298). A quasi-appropriable rent of an asset occurs when it is expensive to remove the asset from the contractual relationship once it has been installed or used as part of the relationship, or it may be so specialized for a particular use that the owner of the asset faces the situation whereby his asset is indeed "locked-in" into the contractual relationship. In such a case, the entire quasi value of the asset (that is, its value in its next best use) could become liable to appropriation by the other contracting party in the case where he or she either chooses or has the chance to behave opportunistically.[6]

Williamson (1983: 526) identifies four different kinds of asset specificity: site specificity refers to situations wherein two dependent stations of economic activity are located right next to each other so as to economize on inventory and transportation expenses;[7] physical asset specificity are situations wherein there is a specific input that is required to produce a product; and human asset specificity arises in cases of learning-by-doing jobs. The fourth kind of asset specificity is called dedicated assets and it refers to investments of a firm or entity relating to its production capacity that would not have been undertaken "*but for the prospect of selling/buying a significant amount of product/input to a particular consumer*" (Williamson, 1983: 52, emphasis added). Dedicated assets, therefore, are put in place as a response to certain kinds of supply agreements and would result in excess capacity that cannot be so easily used, in case these agreements are terminated prematurely (ibid).

A general but decisive feature of all these kinds of asset specificity is that the assets lose value if employed in alternative uses (see, among others, Williamson, 1983; Hart, 1995; Klein et al., 1978).

Transactional factor III: Frequency of a transaction

Frequency of a transaction is a parameter that conditions how parties may choose to decrease transaction costs between themselves where they repeatedly transact with each other. When transactions are frequent, parties will choose to invest in some sort of governance structures that make frequent transactions reliable and cost effective.

Categorizing transaction costs

A more convenient categorization of transaction costs in a way that permits more in-depth analyses was first proposed by Willamson (1971): search and information costs, bargaining and decision costs, and, monitoring, renegotiation and enforcement costs.

Search and information costs

Search and information costs refer to those costs incurred by individuals in order to obtain information about the various options available for conducting a particular transaction and in finding desirable partners with whom to conduct the transaction. Getting informed about the options available and searching for the right partner involves costs that could relate to gathering information, advertising one's own capabilities, or those involved in testing and quality control (Furubotn and Richter, 1998: 44–45). These costs also include obtaining information about the contingencies that may arise during the life of the contract and, in extreme cases, even to get information about the probabilities under which the undesired state of the world may occur.

Bargaining and decision costs

Usually, parties to a contract face costs of negotiating and reaching an agreement over the various clauses contained in the contract, costs of anticipating contingencies, and costs of finding potential solutions. Incorporating all these details into the contract in a written form also entails costs. As Hart (1995: 23) puts it, "the individual plans so made by investing in all relevant information required cannot be negotiated easily amongst the contracting parties, the least of their problems being finding a common language to express their common concerns".

Monitoring, renegotiation and enforcement costs

A contract per se does not guarantee contractual performance at the level of quality expected by the contractual parties. As a result, costs of checking that the other party abides by the contract in a desirable way may itself be prohibitively costly. Since parties do not have all the information about all states of the world that may materialize during the life of the contract when these states of the world do materialize, they have to renegotiate some or most of the terms of the contract. When contractual parties are faced with information about new states of the world with the possibility that one or both of them resort to opportunistic behaviour, *ex post* negotiations may lead to situations of hold-up, especially in the presence of relation-specific investments.

Post-contractual costs can also be incurred when the parties decide to set up their own governance structures for the contract. There may be situations where contractual parties anticipated some of the *ex post* contingencies that could arise, and provided for a third party intervention – for example, a court. Given the possibility that the court or the designated third party cannot verify the dispute in an objective way, parties may incorporate specific means into the contract by which the contract will be amended when the need arises. These "means" or "governance structures" also impose costs of administration and enforcement on the contract (Kreps, 1995: 745).

The resulting contract: Essentially incomplete

As already mentioned, market imperfections make it difficult for individuals to contemplate and contract for every contingency that might arise during the lifetime of a contract. From an *ex ante* perspective, not only do these factors prevent individuals from being able to foresee each and every contingency that may arise during the contract, there are situations where individuals anticipate some of the contingencies but refrain from taking precautions *ex ante* due to the high costs involved in doing so (see the discussion in Kreps, 1995: 744–745). These costs – anticipating the various eventualities that may occur during the lifetime of the relationship; deciding and reaching an agreement about how to deal with such eventualities; writing the contract in a sufficiently clear and unambiguous way such that the terms can always be enforced; renegotiation and enforcement – are the main reason for contractual incompleteness (Hart, 1995).[8] Contracts will therefore have gaps or missing provisions about how to deal with some states of the world when they in fact materialize. These provisions will necessarily have to be renegotiated and revised through contractual negotiations, or, in the worst case scenario, be sorted out through legal settlements (ibid).

The solutions proposed by transaction cost and information economics

Transaction cost economics holds that if the key dimensions of a transaction can be identified (in terms of the standard behavioural assumptions of bounded rationality and opportunism and transactional assumptions of uncertainty, frequency, and asset specificity), efficient institutional arrangements can be predicted (see Holmstroem and Roberts, 1998).[9] The way any given transaction adapts to circumstances as they arise depends on the characteristics of the transaction itself. Each one of these characteristics places demands on the contractual environment in specific ways (in terms of the costs they impose), thereby mandating specific institutional arrangements. For example, when higher degrees of uncertainty and asset specificity occur together, they result in a more complex contractual environment that necessitates parties to make greater adjustments after the contract has begun (Holmstroem and Roberts, 1998: 76). Therefore, institutions are transaction-cost economizing instruments and the significance of alternatives modes of organization of economic activity lies in their relative advantages/disadvantages in helping parties deal with the various transaction costs.

But the major weakness of transaction cost economics is that it only serves as a tool to analyse, or as Furubotn and Richter (1998: 55) call it "at least describe" organizations, actors, and their constraints in transaction cost terms. Modelling transaction costs concretely, quantifying them and their effects, and proposing appropriate solutions has been dealt with in microeconomic literature by including the existence of such costs into the assumptions of other theories and models. This is done by information economics by analysing the consequences of asymmetric information, principal-agent theory and the property rights theory of the firm.

III. SEARCH AND INFORMATION COSTS IN THE MARKET FOR BIOPROSPECTING

The term "search costs" can denote two different kinds of costs depending on whether it is used in the neoclassical economics sense or in the context of information economics. When used in the neoclassical sense, the hypothesis is that, since searching is costly, parties will optimize their search efforts until marginal search costs and marginal benefits are equal. Construed in this way, search costs do not pose a problem to contract formation.

Search costs also denote search and information costs as construed within information economics. These are costs imposed by information asymmetries between people. This is the sense in which the term "search costs" is used in this chapter.

The impact of information asymmetries on contracts: A closer look at information economics

According to information economics, in the absence of complete information, private information hinders parties from reaching efficient agreements or even causes market failures. As a result, in information economics terminology, private information is an incentive constraint. The term "incentive constraint" denotes any parameter that hinders the creation of appropriate incentives for people to enter into wealth maximizing exchanges. Overcoming the problems of contracting with private information requires that parties evolve, what are called, incentive-efficient mechanisms to enable contractual relationships.[10]

Two sets of problems faced by contracting parties are very important – those that are dealt with under the headings of hidden action (or moral hazard)[11] and hidden information (or adverse selection).[12]

This section mainly focuses on problems of adverse selection and the costs of search and information gathering associated with it. Problems of moral hazard are dealt with in Section V on monitoring and enforcement costs.

Information asymmetries and the scope for pre-contractual opportunism

If one party to a contract is more aware of the costs and benefits of the transaction than the other, the likelihood of different possible outcomes may be private information to that party alone. Such informational asymmetries can prevent a contract from being entered into, even if it is efficient to enter into such a contract under a situation of complete information (Milgrom and Roberts, 1996: 140). Information asymmetries create pre-contractual opportunism because they create the potential for misrepresentation of value in order to get a better price than the competitive equilibrium price. Or, even in cases where both parties to the contract are equally aware of the approximate potential value of the good, they may differ in its valuation (ibid). This situation leads to what is called the "winners curse" where the more knowledgeable party allows you to buy or sell something that is worth less than you thought at a high price or vice versa (Zeckhauser, 1996: 12745).

Logically, one would assume that in the case of informational asymmetries, the party who has the additional information will try to extract an "informational rent" from the transaction.[13] But in many cases, informational asymmetries lead to both parties expecting an informational rent from the transaction. Normally the seller has more information on the properties of the good than has the buyer, but the seller, in the absence of accurate information, can presume that his good is worth much more than its actual worth. In such cases, although the buyer values the good much more than the seller does, such an expectation of both parties to extract rents from one another will hinder them from reaching any agreement about the price at all, thereby marring the possibility of exchange.

Such strategic misrepresentation of the quality of the good is more of an acute problem if the good that is to be traded is information itself – which is the case in many bioprospecting contracts. The economic properties of information as a good – that of non-rivalry in consumption and non-exclusiveness[14] – cause acute contractual problems when trading with information as a private good.

But the assumption that information is a non-exclusive commodity does not always hold true. Exclusive monopolies can be created through intellectual property protection where economic rents are guaranteed to the owner when the information is traded. Intellectual property protection can also serve as a signal of quality as it indicates potential of innovativeness. But in the absence of any such mechanism, the problem is as described by Zeckhauser (1996: 12744), "When the good in question is said to have characteristics that cannot be verified on the spot, the buyer faces two open questions: Will the product contain what is being claimed by the seller? And if so, will it facilitate the development of new products?" Apart from such uncertainties about the exchange itself, agreeing upon a price for the information requires that the buyer assesses the demand for the potential product that he can create using the information (ibid.).

Mostly, unless the seller can signal quality of the good, the problem of uncertain product quality in the market leads to adverse selection and, in extreme cases, to a "market for lemons" where only low quality goods will be traded (Akerlof, 1970).

Solutions to the problem of adverse selection

Models of information economics have developed screening, signalling and reputation effects as mechanisms to prevent contractual distortions caused by adverse selection and moral hazard issues.

Screening refers to the process of designing contractual provisions in such a way that only the desired category of people are attracted (Milgrom and Roberts, 1996). For example, *auction mechanisms* are a way of ensuring that only the desired category of buyers are attracted ("screened") for the sale of a good. Auction procedures are tailor-made for situations where the person who has something to sell has no idea about the willingness of parties to pay for the item, but would like to receive the highest revenue/utility from such a sale (Myerson, 1981). The auction mechanism, the rule by which the winner of an auction is determined, has to ensure that potential buyers reveal to the seller their true valuation of the good instead of strategically under-representing it.

Signalling means that the party that possesses private information takes steps, which when properly interpreted, reveals this private information (Spier, 1992). For example, a person's educational attainment (grades/university attended) provides a signal of intellectual capability to a potential employer who otherwise could only assess that person's capability after assessing their work over a period of time (Milgrom and Roberts, 1996).

Reputation effects refers to the significant benefits of a "good" reputation in markets with repeated transactions, where assertion of quality of the good is a problem. When there is a scope for repeated transactions, the building of "goodwill" is a signal of transaction worthiness of parties. The solution of reputation effects has a lot in common with what is called relational contracting in the literature. Relational contracting refers to cases where parties deal with each other repeatedly and therefore the perks of having a long-lasting contractual relationship with the same party leads to cooperation even in cases of conflict.[15] Hence, reputation effects have been proposed as a mechanism for motivating mutual commitments amongst parties in incomplete contracts.[16]

How the solutions of screening, auction mechanisms, signalling and reputation effects can be made to work in the market for bioprospecting is discussed in Chapter 6.

Traditional knowledge and genetic material transfer: A market for lemons?

The problem of uncertain product quality in the market for traditional knowledge-related information and tangible genetic resources, where the quality of genetic resource samples or the reliability of the traditional knowledge-based information or both can only be verified after the good has been traded, leads to problems of adverse selection and, in extreme cases, to a breakdown of market exchange.

Spot-market practices for the sale of tangible genetic resources and related information is common. There have also been frequent reports of traditional knowledge-related information being traded single-handedly in the literature. Where traders of traditional knowledge are concerned, misrepresentation of both the characteristics of plants, and the quality and amount of traditional knowledge possessed is possible, since the firms/access-seekers can only screen in the immediate future; the remainder of the medical research on particular characteristics and compounds can take years to achieve. As a result, uncertainty about quality prevails in these transactions. As Laird (1993: 112) observes, "there is a tendency to overstate the importance of the genetic collections to pharmaceutical firms that launch such research and development programs and hence demand large sums of money that has already incensed many in the pharmaceutical industry."

Cunningham (1996) notes the reaction of some pharmaceutical companies who, in order to overcome this problem somehow and propel the market in a direction where reputation effects can play a role, have set out their policies for acquisition of plant materials in the search for active ingredients as a result of uncertainty about quality. Glaxo Group Research Ltd. (now Glaxo Smithkline), for example, is reported to have launched a policy to conclude agreements with bona fide organizations only rather than with individuals. This policy also includes covering costs of supply and recognizing the negative effects that unauthorized material can have on the environment and economies of the countries concerned, particularly in the developing world (ibid.).

But to get rid of product uncertainty caused by information asymmetries completely would necessitate the optimum assessment of sample quality. Zeckhauser (1996: 12744) sums up this problem aptly: "Just displaying the information/the product that contains all this information is not sufficient to convince the buyer to purchase it, but communicating to him about the product and its qualities will already amount to revealing the information that he proposes to sell. As rightly noted, sometimes even to make clear something is technologically feasible is already giving away a lot; it reveals not only that innovation is possible, but also that the effort is worth trying."

The problem of ascertaining quality of genetic resources and traditional knowledge-based information is less of a problem in the case of pharmaceutical drugs than in botanical medicines. Since botanical medicines are in many cases based directly on the traditional medicinal use, and also rely on the traditional medicinal efficacy as the selling point (Laird, 2004, personal communication), just revealing the information may give away a lot.

One has to distinguish between two distinct yet related problems here. The first is when firms really have misrepresented the value of the end product or the extent of usage of traditional knowledge to communities and access authorities and subsequently purposefully failed to disclose the true value of such inputs to the research process. This kind of behaviour would fall into the "biopiracy" category.

The other problem here could be that, as a result of informational asymmetries, both parties to expect an informational rent from the transaction. This problem is especially significant in the market for bioprospecting because, based on earlier examples of biopiracy, the communities and the national access authorities tend to mistrust the extent of information given out by the firms/users as well as believe that their resources are extremely lucrative to the firms, which may or may not be the case. The firms, on the other hand, would possibly like to rid themselves of the badge of "biopiracy" and engage in mutually productive collaborations, but uncertainty related to the R&D process prevents them from making very high *ex ante* benefit-sharing promises. In the absence of mechanisms that could rectify the lack of divulgence amongst the parties about the research goals and progress that has seemingly led to the collapse of bargains or even to adverse publicity in earlier bioprospecting contracts, this problem is unlikely to be resolved.

A second and more important impact of this sort of pre-contractual opportunism is that parties prefer to rely on the market-based intermediary structures wherein smaller firms or other intermediary organizations are involved. This mode of contractual exchange ensures, through the mechanism of reputation effects, that only good quality tangible resource components or traditional knowledge-based information are exchanged. Intermediaries such as in-country suppliers, botanical gardens, or even small dedicated firms that wish to operate within a given niche in the biotechnological industry, have a reputation to safeguard since, on the one hand, they will indulge in repeated transactions with drug firms and, on the other hand, with source countries.

But reputation effects will not be disciplining factors in the case of large drug R&D programmes based on genetic resources, since there is very little scope for repeated transactions in the sense that similar programmes will be initiated between firms and the same source countries and/or communities.

IV. BARGAINING AND DECISION COSTS

Bargaining and decision costs exist to a certain extent in all contracts. Only when they become significant, do they have the effect of rendering a contract incomplete. Two forms of uncertainty in bioprospecting – process-related uncertainty and legal uncertainty – impose high costs of bargaining and reaching mutually acceptable decisions over the terms of the contract.

As already discussed in the section on contractual incompleteness, implementing a contract under uncertainty requires that parties agree to renegotiate when unforeseen contingencies materialize. But problems in renegotiation arise if for any given new state of the world (a) some parties were to realize benefits by declaring it a false state of the world (opportunism) or (b) information regarding the state of the world is dispersed unequally among the parties (information asymmetries) or (c) despite possession of identical information by all parties, agreement on how to deal with this new contingency must still be reached (negotiation costs) (Williamson, 1975: 101).

In the light of such uncertainty, when one party to the contract deviates, it is: (a) extremely costly to enforce desired behaviour, and, more importantly, (b) difficult to verify facts in front of a third party, for example, the courts. As a result, parties face problems in negotiating the contracts in the first place, because obtaining credible commitments is difficult or the knowledge that they would have to revisit certain key terms within research collaborations at each stage of the contract acts as a deterrent.

Process-oriented uncertainty

The extent of uncertainty in the drug R&D process based on genetic resources has been already been discussed in this book. When samples are chosen for drug R&D through bioprospecting, the future is absolutely unforeseeable to all parties involved – none of them can actually predict with certainty whether a drug will be discovered at all. Even if the parties behave rationally, their information about the future is imperfect. As a result, the contract can at best be only incomplete where important gaps, specifically those related to benefit-sharing, will either be imprecise (if decided upon *ex ante*) or can only be filled in after breakthroughs at each stage of the drug R&D process. Due to these problems in price-setting, two good solutions may be that (a) the parties agree on an up-front payment, but also be clear on the fact that a very crucial aspect of the contract will be renegotiated at a later date when more information on R&D success or failure becomes available, or (b) settle upon mile-stone payments. Milestone payments refer to payment schemes where several milestones in the drug R&D process are identified, and benefit-sharing instalments are set conditional on the process reaching those stages. In such payment schemes, communities or national governments receive incremental benefit-sharing packages as and when the R&D progresses from one stage to another.

Legal uncertainty and its impact

Legal uncertainty refers to any or all of the three kinds of situations: (a) where there are no laws to govern access to genetic resources and use of traditional knowledge; (b) where there are laws, but they do not clearly provide for the roles and responsibilities of users and providers; and, (c) where the political and legal situations within the country are unstable, as a result of which a "new" legal situation may emerge at any given point of time. All these kinds of legal uncertainty have the potential of rendering contracts incomplete that could otherwise have been complete.

Such situations act as a disincentive for firms to invest in R&D based on genetic resources and/or traditional knowledge. The fear that the agreed actions may not be carried out, or the fear of getting cheated (whether parties really do it or not), may prevent an efficient transaction from taking place (Milgrom and Roberts, 1996: 133). Uncertainty amongst companies about the status of their research when they use genetic material obtained without permits may prompt them not to take the risk of using such samples in their R&D processes at all (ten Kate and Laird, 1999: 298–301). Companies are also wary of working with source countries that have very stringent or extremely elaborate access and traditional knowledge procedures (ibid.). Problems caused by such uncertainty can only be solved by extensive stipulations within the contracts, which in turn can only be achieved with a lot of costly legal expertise. This is also part of the reason why companies prefer to have other intermediaries who deal with these cumbersome procedures of obtaining access and collecting useful genetic resources.

In the absence of legal certainty, it is a formidable task to extract credible commitments from contracting parties. The possibility that certain claims of profit agreements are not enforceable in courts, or that one of the parties lacks the technical knowledge to prove their claims in the courts even when contested, leads to opportunistic behaviour.

The fallout of ex post *negotiations*

In situations of uncertainty, both process-oriented or legal, despite the fact that it may be beneficial to all parties in bioprospecting to renegotiate the contract at each stage based on new information generated through the R&D process, such renegotiations may not take place due to several reasons.

First of all, if the scope for such renegotiation is clear to both parties at the time of entering the contract, it may hinder them from drafting the contract in a way that generates the desired incentives. Secondly, to the extent that the parties possess different levels of information and that there exists immense legal uncertainty, the fear of getting cheated in the *ex post* bargaining stages may result in a total lack of unanimity about any sort of contractual solution (Hart, 1995: 25). Lastly, *ex post* negotiations can be hampered from proceeding in a productive way even under legal certainty if there are political influences that play a role in the negotiations. These could also be called political costs, as Furubotn and Richter (1998) categorize them, and refer to costs of two kinds. They can either be the costs of rule-making which, in the case of bioprospecting, means the costs related to reaching international consensus on the broad features of the rules that ought to govern access to genetic resources and benefit-sharing. This will also include the costs of rule-making at the national level. The second kind of cost is the cost of running the already established rules. This can also be significant because the institutions required to enforce the rules may themselves be costly, or the different interests groups permeate even within national levels and can offer resistance to implementing national frameworks.

Both categories of costs imposed by political influences can be a real threat in the case of bioprospecting collaborations. Given the diverse stakeholder interests, there is immense reluctance among industry partners to deal with the cultural and politically sensitive arrangements within source countries (Rosenthal, 1997). For example, in the case of the Suriname ICBG programme, the contractual arrangement specifically separates the culturally and politically sensitive arrangements with local or indigenous providers of tangible and intangible resources from the commercial research agreements that may include industrial partners. According to Rosenthal (1997: 267), the reason for this, in part is "the reluctance some industrial partners have shown in negotiating directly with local community groups that are often remote, and may have both unstable organizational structures and little understanding of the commercial and legal environment of industrial research". Similarly, other authors have noted that one of the reasons for the Merck–INBio contract's success is partly because indigenous knowledge was not a part of the contract (see review in Mann, 1996).

Two other ICBG programmes – the ICBG Peru and ICBG Maya cases – are also good examples to demonstrate the impact of uncertainty in bioprospecting contracts. The ICBG project in Peru consists of several collaborative arrangements between the five partners listed in Chapter 2. Two separate contractual negotiations – between Washington University that got the ICBG grant in 1993 and Consejo Aguaruna and Huambisa (CAH) through a letter of intent, and a separate contract between Washington University and Searle (for product development) are of specific interest here (Tobin, 2002). After failure of negotiations between the CAH and Washington University, a contract is now in place between the latter and three local Aguaruna peoples' federations for supply of medicinal plants that are in use by their communities. Searle, on the other hand, sought to negotiate for a warranty from the Aguarunas that their knowledge was free of third party rights. In the event of third party claims, Searle wanted a declaration by the Aguarunas that they forfeit all royalties and that Searle was free to use the knowledge as it chose fit (Bell, 1997; Tobin, 2002: 307). This was not considered to be favourable for the communities, since claims by another community in the region on the knowledge components used by Searle would result in all of them losing royalties and rights on how such knowledge should be put to use. Given the absence of a law in Peru that dealt with issues of access to genetic resources and community rights in a comprehensive way, other ways of resolving the conflict were sought. A "know-how" licence now operates between Searle and the Aguarunas, as a measure to prevent the company from gaining rights over their knowledge and tangible resources which would infringe upon the Aguarunas' own collective property rights. Given that the Peruvian government has all rights on the genetic resources within the country, the "know-how" requirement ties the rights to use the plants with the need for a licence to use the knowledge provided by the Aguarunas (Lewis et al., 1999). Searle now pays them annual licence fees, as well as milestone payments (ibid.). It was also settled that if the agreement is terminated at any point, Searle's rights to use the plants' derivatives will be lost (Bell, 1997).

The ICBG Maya case, on the other hand, is a very interesting example of how legal uncertainty can create a situation where *ex post* negotiations are completely taken over by political influences.

The International Cooperative Biodiversity Group programme in the Highland region of Chiapas, Mexico

The Mayan ICBG was one of the International Cooperative Biodiversity Group programmes that received a grant in 1993 to run a drug discovery programme based on medical ethnobiology of the Mayan Communities in the Highland region of Chiapas, Mexico. It was aimed at: researching pharmacologically important plant species found in the state of Chiapas; finding potential treatments for diarrhoea, respiratory conditions, contraception, and other illnesses of local importance; initiating a comprehensive ecological survey of the vascular flora of the Highlands of Chiapas and enhancing the research infrastructure at ECOSUR. The programme began in 1993, under the leadership of Professor Brent Berlin, an ethnobiologist from the University of Georgia. It had to be halted in October 2001, when the host country institution, El Colegio de la Frontera Sur (ECOSUR), withdrew its support in a highly politicized atmosphere of mistrust amongst the various contractual parties.

The tension amongst the various contractual parties started when a group of traditional healers, midwives, and activists from national and international organizations led by RAFI (the Rural Agriculture Foundation International, now called the Action Group on Erosion, Technology and Concentration) mounted accusations on the legality of the agreements on the collection of genetic resources and ethnobotanical knowledge as part of the programme. Some of the main issues that formed the basis of the dispute were:

(a) How to define Mayan ethnobotanical knowledge? Was it knowledge that belonged only to specialists and traditional Mayan healers (organized under groups such as the OMIECH and the Traditional Healers Council of Chiapas) or did it belong to the Mayan communities?

(b) Who has control over the natural resources within Mexico? Although Articles 87 and 87 Bis of the General Law of Ecological Equilibrium and Environmental Protection deem all natural resources to be the sole property of Mexico and that such collections should only proceed with legal approval, there is no specified procedure on how prior informed consent is to be obtained. Although collections under the programme were initiated with a properly issued permit in four municipalities, the legitimacy of the consent procedure was questioned by pressure groups.

(c) How can equitable benefit-sharing be designed when tangible genetic resources and intangible ethnobotanical knowledge over their use were held by different parties? In such a case, whose claim is more superior and what is the procedure for obtaining prior informed consent for both?

Attempts by the researchers of the ICBG group to negotiate solutions to the dispute were not successful due to several reasons. The lack of political and social empowerment of the Mayan people affected the dispute settlement process tremendously. It was not possible for the other Mayan communities in the region (other than the healers' groups and those that supported the pressure groups) to organize themselves in response, because of lack of empowerment amongst them, since there is no formal structure of representation amongst the Mayan communities in the Chiapas region. The situation worsened in 2000 when the then elected Mexican government decided that the Mexican bioprospecting law needed review. This, coupled with the legitimacy claims being made on the prior informed consent obtained by the ICBG Maya group, led to a cessation of collection activities in the region in 2000. The efforts by the National Institute of Health and the ICBG Maya Group to focus only on the prior informed consent procedures to solve some of the controversial issues were also not encouraged and supported by ECOSUR due to growing internal pressures within the country from interest groups such as non-governmental organizations. Eventually, as a result, the project came to a halt in 2001. The ICBGs have themselves been the subject of many critical reviews since 1996, when USAID withdrew its support for the programme. Halting the drug discovery programme not only put an end to exploring the potential of Mayan ethnobotanical knowledge for modern drug discovery within the ICBG framework, but also created an uneasy precedent for future bioprospecting. Mexico's biodiversity landscape is torn between the politics of land rights and autonomy issues amongst the indigenous groups and the government, which in this case, proved to be a fertile landscape for an escalation of conflict by non-governmental organizations. Unfortunately, the reputation of a dedicated research ethnobiologist and his partner who had been committed to the cause of ethnobotanical research since the 1960s was also caught up in all this.

Sources: Brent Berlin (2004), personal interview; Dalton (2001); Hayden (2003).

V. MONITORING AND ENFORCEMENT COSTS

Whenever the terms of remuneration are set in a way that contractual performance affects one party more than the other, there is a risk that the party who benefits less from contractual performance in the *ex post* scenario does not have sufficient incentives to abide by what was originally agreed upon. Economic literature terms such behaviour "moral hazard". Moral hazard is a common contractual problem, although it has its origin in the economic analysis of insurance contracts. The terms on which an insurance contract is concluded bears a large impact on the *ex post* behaviour of the insuree. For example, total accident insurance provides the insuree with incentives to indulge in more risky behaviour. Health insurance creates incentives for insurees to indulge in practices that may cause adverse health impacts – that is, it may encourage practices that insurees would not have indulged in had they borne the full costs of illness. Similar monitoring and enforcement problems exist in other forms of contracts too – for example, in corporate contracts, managing directors of corporations who hold few or no shares will not have adequate incentives to devote their best efforts to increasing share value, since they will not benefit from such increase as much as the shareholders.

Principal-agent theory proposes *ex ante* incentive alignment as a potential solution for such problems. *Ex ante* incentive alignment refers to the process of designing incentives in the *ex ante* stage so that such divergences are minimized. For example, insurance contracts that only cover part of accident damages or limit insurance coverage to specific instances have the effect of automatically placing a check on the behaviour of the insuree. Similarly, divergence between interests of share holders and managing directors within corporations can be reduced by devising renumeration schemes that make it more attractive for managers to pursue the best interests of the shareholders.

Where *ex ante* incentive alignment is not possible and parties have access to private information during the course of implementation of the contract, monitoring issues will persist due to the potential for opportunistic behaviour on the part of one or both contractual parties.

Monitoring problems in bioprospecting

There are three separate principal-agent relationships in bioprospecting – the first is between the firms which contract for supply of genetic samples and/or traditional knowledge and the national access authority, the second is between the access authorities and the local and indigenous communities whom they seek to represent, and, the third is between the national access authorities and the firms in the *ex post* contract stages where the firms have more information than the national authorities about the progress being made in the R&D process, but have systematic incentives not to reveal them.

From a firm's perspective, the main question is: is it easy for a firm to monitor the efforts of its agent (the access authority, the communities, or any other intermediary such as smaller firms) in the post-contractual phase? For the firms, it is easier to monitor the quality of the supplies when they choose freelance intermediary suppliers as this allows the supply to be terminated whenever control over sample quality cannot be exerted. This solution may be preferable to long-term contracts which cannot be terminated easily. This way, they can depend on monitoring by markets to a limited degree. But the market is quite specialized and often, even while operating on a freelance basis, there is still a need to fix benefit-sharing structures such that suppliers can promise quality along with quantity. For this, firms prefer to use incentive contracting as a mechanism, which may go against the basic risk preferences of parties in bioprospecting contracts, as well as increase the monitoring costs of access authorities in source countries in the *ex post* contrast stages. The ramifications of this trade-off are explained in detail here.

Parties can exhibit different attitudes towards risk bearing depending on their preferences for certain incomes over uncertain incomes with the same expected value. These preferences can be characterized by three different formal categories. A person who prefers a certain income to an uncertain one with the same expected value is described as *risk-averse.* Risk aversion is said to be the most common preference towards risk (Mas-Colell et al., 1995: 449). In contrast, a *risk-neutral* person is one who is indifferent to all alternatives with the same expected value (Katz and Rosen, 1998: 168). *Risk-loving* refers to the tendency where an uncertain value is preferred to a certain value of the same expected value (ibid.).

Applying this categorization to bioprospecting, the collectors of plant samples (and possibly, associated traditional knowledge information) are more risk-averse than the drug firms. Collectors want their salary fixed to the extent that their interest in basing their income levels on the probability of drug development using their samples are very low. This would hold especially true in the case of in-country collectors or organizations since their ability to monitor the flow of genetic material after it leaves the national frontiers is very limited. If the holders of the knowledge (the communities) are assumed to be the contracting parties instead of the collectors, it is still probable that the communities are more risk-averse than the companies given their low income levels; and therefore they will prefer higher certain incomes with lower shares of uncertain benefits, rather than lower certain incomes with high-powered benefit-sharing of uncertain benefits in an uncertain future. The communities can be assumed to be risk-averse because of their low levels of income and the lack of possibilities to diversify their income sources. Mistrust caused by instances of biopiracy may also cause risk-aversion amongst communities and their representatives, the national access authorities, as a result of which they prefer to receive upfront payments. Since usually genetic resources pass through a chain of intermediary organizations in the drug R&D process, source countries (represented by access authorities) also prefer large upfront payments and smaller reliance on royalties due to problems of monitoring. The intermediary organizations, such as the dedicated biotechnology firms also may prefer to have fixed upfront payments rather than royalty payments. Laird (1993: 153), in a survey of small and medium-sized firms who offer such services, notes that firms prefer high upfront payments and the reliance on royalty is smaller (1–3%). And finally, the large drug firms as well as government initiated collaboration projects, although risk-averse themselves, are in a better position to diversify risk by carrying out the activity simultaneously in different source countries.

Given the risk inherent in joint research efforts for drug R&D, the result can be one of high income in the good state of the world (the knowledge pool/genetic resource collection is of high quality and leads to marketable products) or low income in bad states of the world (bad quality knowledge pools or genetic resource collections with no products). Economic logic would predict that since parties are bound to have different attitudes to risk-bearing, the contract design should ideally ensure optimal risk-sharing amongst parties, thereby maximizing their expected utility from monetary incomes. This will be achieved in an *ex ante* perspective by lowering the income level of the good outcomes and redistributing it in favour of the bad outcomes (Katz, 1989).

But a survey of older and ongoing contract structures reveal that due to the preference for incentive-efficient contracting by drug firms, the different attitudes of other actors towards risk are not reflected in the contractual structures in bioprospecting. For example, Aylward (1995) notes that the in-country collectors receive an initial payment of £25 and 50 per cent of the eventual royalties obtained by Biotics from the pharmaceutical industry in case the screened compounds result in successful drugs at a later date (Aylward, 1995: 116). Although the actual terms of these contracts are mostly confidential, it is clear that the raw materials gain in value as they move up the R&D chain. For example, although Biotics paid in-country collectors around £25 for the supply of one sort of raw material plus 50 per cent of the royalties that Biotics itself was to receive, average figures for the industry reveal that Biotics received an initial payment of roughly US$50–$200 on delivery of samples to the pharmaceutical partners (Aylward, 1995: 117; Laird, 1993). Many other similar contractual arrangements can be listed where the reliance is on lower up-front payments and higher royalty rates (i.e. higher royalty rates in a relative sense, when compared to the risk preference of the contracting party).

The same preference for incentive contracting is also evident in bioprospecting collaborations. For example, in the ICBG Suriname Project, all benefits are split up between advance payments and royalties to be paid to the Forest Peoples' Fund and five other governmental and non-governmental organizations, many of which represent the communities who will supply ethnomedical inputs for the research process (Rosenthal, 1997: 261).

The fallout of incentive contracting: The growing issue of monitoring for governments

Opting for contractual arrangements which prioritize incentive contracting over the risk preferences of the various parties no doubt reduces post-contractual opportunism of the communities and the national access authorities and is therefore more efficient for the firms. But this increases the scope for post-contractual opportunism on the part of the firms since they are the parties to the contract which are more likely to have precise information about the success of the R&D programme and its commercial prospects at each stage. This has unleashed a range of concerns on monitoring the flow of genetic resources and traditional knowledge-related information amongst source country governments and organizations.

Monitoring is a crucial element of regulation of access in itself and of assuming guardianship of traditional knowledge because, in the absence of being able to follow the value increase of the genetic resources and the associated traditional knowledge component, ensuring "appropriate" compensation is almost impossible. The main issue for the access authority is to be able to monitor (a) the flow of the genetic resources and associated traditional knowledge along the drug R&D chain, and (b) the proportional increase in its economic value. How can the access authority be sure that the firm seeking access will not cheat/or misrepresent the potential value of the leads that are being derived out of the samples collected *ex post*? When the final product stage is reached and the firm claims that it was not from the samples collected from country X where the firm performed extensive collection operations, can the access authority be sure of this fact?

The fact that many bioprospecting cases are transnational in nature makes these questions of monitoring more urgent for source country governments and communities. Concerns have been expressed that in a case where the firm refuses to reveal the details of the drug research, it would be impossible for the national access authorities or for the communities to get such information to prove details of the R&D programme to a court or a different third party. Noting the same in the context of the Merck-INBio contract, Mooney (2000: 40) argues that "If at some stage Merck argues that its latest wonder drug either sprung unnaturally from a molecular model – or from the tropical archives of a Bronx herbarium – how will INBio or Costa Rica argue otherwise?"

This fear of misappropriation of genetic resources and traditional knowledge has led to very stringent laws at the national level. The negotiations on including certification of origin into patenting requirements that have been discussed in Chapter 3 may help solve this issue. To what extent this can be a solution to monitoring problems is discussed in Chapter 6.

VI. IMPERFECT COMMITMENTS AND SPECIFIC ASSETS

The classical "hold-up" problem that was first enumerated by Klein, Crawford, and Alchian (1978) has served as the starting point for the literature on asset specificity in incomplete contracts. Two different theories have evolved over the past two decades on the impact of hold-ups on contractual relations and how this problem can potentially be resolved. The theory of transaction cost economics, largely dominated by Williamson's works, predicts vertical integration as the solution to the problem of hold-up (see for example, Williamson, 1975, 2000, amongst others). The property rights theory of the firm *or* the incomplete contracting theory of the firm, on the other hand, presents some of the state-of-the-art analysis of issues of asset specificity and its impact on the nature of ownership structures.[17] The property rights theory of the firm is an attempt to develop rigorous foundations for the hold-up problem and integration decisions by firms. Since this theory has its foundation in the works of economists Grossman, Hart, and Moore, it is also referred to as the GHM framework (see for example, Williamson, 2000).

Both these theories have similarities: they both deal with decisions of investing or not investing when there is bounded rationality (or uncertainty in the case of GHM), when parties to the contract are bilaterally dependent due to asset specificity, and, lastly, non-contractibility cannot be easily overcome in either theories due to opportunistic behaviour (Williamson, 2000: 605). But to a very large extent, the property rights theory of the firm can be seen as a reaction to theoretical shortcomings in the transaction costs economics framework.[18]

Asset specificity and the theory of transaction cost economics

Coase was the first to explore at length the underlying causes that made firms more attractive than markets to organize specific forms of economic activity (Coase, 1937). According to Coase, firms arise because organizing production within a firm through hierarchies would be cheaper than coordinating the production process through markets with independent firms on a contractual basis. He identified the main cost of using the market to be the costs associated with haggling and getting informed over the terms of trade. These costs can be significantly large in the case of long-term transactions where learning and haggling may need to be done several times. Organizing such transactions through a firm allows one party to have a higher authority over the terms of trade, thereby reducing such costs of organizing through the market (see Coase, 1937).

But although Coase was the first to identify the presence of these costs, the nature of these transaction costs remained unclear, as a result of which the "benefits" of organizing a given transaction within a firm were quite vague (Grossman and Hart, 1986: 692). Williamson conceptualized and codified these transaction costs for the first time. As already discussed in the earlier sections, transaction cost economics identifies asset specificity as a transactional characteristic that creates costs. A contractual relation that is characterized by either high levels of uncertainty or high degrees of asset specificity, or both, leads to *ex post* bargaining costs due to opportunistic behaviour by one or both parties. Therefore, anticipation of a potential hold-up due to unequal bargaining power in the *ex post* stages negatively influences *ex ante* investment incentives of contracting parties, and results in welfare losses (Hart, 1989: 1763). Williamson proposes two efficient solutions to mitigate *ex post* opportunism and to improve investment: (a) parties write a complete long-term contract, or, in the absence of this, (b) parties vertically integrate, where the organization and the governance structure of the firm deal with the issue of hold up (Holmstroem and Roberts, 1998: 74).

In Williamson's framework, if each firm represents one asset of its own, and if these assets have an interdependent relationship, vertical integration always reduces transaction costs. Specifically, if A and B are the two parties in question, they should be able to make themselves better off by the following arrangement. A buys B's asset on which A's asset is dependent and makes B a manager of the new subsidiary. A sets a transfer price between the subsidiary and his own firm equal to the contract price that existed when the two firms were separate assets, and A gives the manager, B, a compensation package equal to the profit of the subsidiary (Williamson, 2000: 605–606). In this way, ownership incentive through vertical integration presents a solution to deal with contractual execution problems that can arise between two independent parties within a bilaterally dependent relationship. In such an arrangement, although the high-powered incentives of a market-based framework are sacrificed for the coordinated management of the assets in a hierarchy with lower-powered incentive structures and bureaucratic costs (of maintaining the hierarchy), the execution problems are completely eliminated (Williamson, 2000: 606).

Despite this significant contribution, there are some weaknesses in the theory of transaction cost economics that have paved the way for further exploration of asset-specificity related issues. Transaction cost economics predicts that when there is asset specificity and uncertainty, parties will vertically integrate. In practice, this means that the complete assets of both firms will merge into one firm as soon as there is interdependency of assets. But there is no cost-benefit calculus of vertical integration in the theory as a result of which several critical issues remain unanswered. When is vertical integration the most efficient solution? And when should vertical integration end? If there are always gains to be had by vertically integrating, how and under what circumstances is integration less viable than non-integration? In the above example, it would always be possible for firm A to buy out any firm B, C, and so on, until the entire production of any economy becomes organized within one single firm.

The theory of transaction cost economics is unable to answer such questions because of several reasons. To begin with, Williamson's theory does not have a robust, quantifiable interaction between uncertainty, asset specificity, and *ex ante* investments. The theory considers the problem of hold up as presented by Klein, Crawford, and Alchian (1978), which assumes contractible *ex ante* investments amongst parties. *Ex ante* contractible investments imply that investments will be made at efficient levels, irrespective of ownership patterns through bargaining (Holmstroem and Roberts, 1998: 77). Williamson, in contrast, argues for the need for ownership as an incentive to invest and proposes vertical integration to deal with asset specificity and uncertainty, thereby implicitly assuming non-contractibility. But he remains completely silent on the nature of such non-contractibility at the *ex ante* stage and how uncertainty causes it. As a result, the theory is inadequate to explain the precise problems amongst parties that vertical integration solves.

Logically, vertical integration should end when reduction of transaction costs through integrating equal bureaucratic costs of a hierarchy. But Williamson leaves the exposition of bureaucratic costs that arise in the process of integration unclear (their nature, extent, and characteristics are all in the open). As a result, the circumstances under which the gains of vertical integration would cease to outdo the bureaucratic costs of organizing transaction within firm boundaries cannot be clearly identified.

A reason for the neglect of the costs of vertical integration also lies in the fact that his approach largely occupies itself with the aggregate levels of asset specificity within relationships and tends to ignore the marginal properties of such asset specificity on the contractual partners (see Hart, 1989; Holmstroem and Roberts, 1998). Ignoring marginal properties of specificity between assets makes it difficult to model integration decisions by firms as optimizing behaviour. As a result, it is not possible to consider the trade off between the costs due to asset specificity (that would motivate firms to integrate in the first place) and the costs of integration itself (that would at some point set the limit on further integration).

All in all, as Hart (1989) notes, although the transaction cost economics approach to the problem of asset specificity and vertical integration throws light on contractual failure, it does not explain how bringing a transaction into the firm would mitigate this failure in a rigorous manner.[19]

Property rights theory of the firm[20]

In contrast, the property rights theory of the firm has been lauded for its dual focus on the costs as well as benefits of integration (see, for example, Holmstroem and Roberts, 1998: 77). The GHM framework takes off from the assumptions of transaction cost economics: it treats ownership of non-human assets as the defining factors of firm boundaries where ownership is seen as an incentive to invest in relationships in which investments would otherwise be deterred by the potential of hold-up problems. But an important point of departure is that the GHM framework tries to explain vertical integration as an optimizing decision based on the relative costs and benefits of vertical integration.

The main assumptions

The focus is on the definition of a firm, ownership, and incomplete contracts as building blocks of the property rights theory of the firm. A firm consists of those non-human assets that it owns or over which it has control. A firm is thus a set of assets under common ownership. That is, if two different assets have one owner, it is a single, integrated firm. If the two assets have two owners, then they are two firms and the dealings between them are market transactions (Grossman and Hart, 1986: 694). Ownership is identified by residual rights of control. Residual rights of control are the rights to determine what can be done with an asset when circumstances arise that are not specified (or were not foreseen) in the initial contract between the parties (ibid.: 692).

As a result, decisions on asset ownership are the main determinants of firm size and boundaries. Decisions on asset ownership (and firm boundaries) are critical because it is the assets that give the owner the bargaining power to influence outcomes when situations unforeseen in contracts arise and when parties have to negotiate how to continue their relationship. Because the owner of an asset is the only one who can decide how to use it, assets become bargaining levers that determine the future payoffs of investing in a new relationship (Holmstroem and Roberts, 1998: 77).

The GHM framework identifies four main costs that cause contractual incompleteness (Hart and Moore, 1998: 1–2). These are: the costs of foreseeing the kinds of circumstances that may arise during the lifetime of the contract; the costs of incorporating all the contingencies into the agreement even when such circumstances can be envisaged; the costs imposed by the conceptualization of appropriate solutions or amendments to the original agreement to deal with all contingencies in case they arise; and, lastly, the costs of legal verification in front of third parties (ibid.).

The GHM framework does not emphasize that contracts are totally incomplete – the idea is more that optimal contracts are partially incomplete depending on parties' abilities to describe the nature of the trade (Hart and Moore, 1998: 3). Furthermore, incomplete contracts are not necessarily long-term contracts. Even in short-term contracts, the same problems of lack of foreseeability and associated costs can arise.

In the case of incomplete contracts that require parties to make asset specific investments, the parties cannot *ex ante* define a division of surplus from their investments. This division has to be determined *ex post* in a non-competitive bargaining situation. The bargaining power of the parties, in turn, is dependent on the control of assets that are chosen *ex ante*. Since the value of such investments is invariably higher within the relationship than outside, parties get "locked in". The property rights theory of the firm defines control as the allocation of residual control rights (ownership) of assets among parties. A transaction within a firm is determined by the owner of the firm's assets. In contrast, transaction or exchange within a market is decided by competitive bargaining. These two control mechanisms result in completely different divisions of surplus from a production process in which many assets are involved.

From an *ex ante* perspective, the different divisions of surplus that are dependent on the asset ownership structure provide higher or lower incentives for the parties to invest into relation-specific assets. Optimal *ex ante* investments, in turn, are necessary to maximize *ex post* joint profits of production. *Ex ante* control over assets (or ownership the way GHM define it) is, therefore, a major determinant of *ex post* optimal production decisions.

The gains from integrating more assets into one firm are that there will be fewer hold-ups amongst parties because when integrated, fewer asset owners have to bargain about *ex post* divisions of surplus. Bargaining thus creates an inefficient externality between owners. Parties without ownership (or reduced ownership) of assets have lower incentives to invest in relation-specific investment that are necessary to maximize joint profits – these are the costs of integration.

The optimal ownership structure of assets balances the gains from fewer hold-ups and the losses from reduced ownership among contractual parties, thereby maximizing the joint surplus from production.

The story of the GHM model

Assume a set of economic agents, $i = \{1, 2, 3, \ldots, I\}$. The agents will have different roles in a joint production process, like workers, managers, or owners of a firm. These agents control a set of assets, $A = \{a_1, a_2, a_3 \ldots, a_n\}$. A firm is a set of agents and the set of assets they control/own. The production process extends over two periods, date 0 and date 1.

At date 0, agent *i* has to decide on the ownership structure of assets a_j and contractible aspects of production, given the information available at date 0. The agent *i* also has to decide on his or her specific investments, that will enhance the productivity of the agent. The gains from their investment can only be realized by a firm or another agent if they cooperate with agent *i* at date 1.

At date 1, new information becomes available that resolves the uncertainty prevailing at date 0 and the production of goods occurs conditional on the specific investments made at date 0. Because of the uncertainty at date 0 about market conditions at date 1, contracts defining optimal investment decisions between agents cannot be written at date 0. Neither is a definition of division of surplus for date 1 possible in a contract written at date 0. The uncertainty about the precise market conditions is resolved only on date 1 and parties have to decide on production of goods contingent on their specific investment decisions at date 0. That is, they will maximize joint profit under the constraint of their investment decisions at date 0.

Bargaining on date 1 will take place under symmetric information and the division of the surplus from joint production will be determined by cooperative bargaining.

Based on the assumptions of incomplete contracts, specific investments, the definition of a firm as a collection of non-human assets and the sequence of investment and production decisions at date 0 and date 1, the property rights theory of the firm develops complex mathematical game theoretic models to derive results about optimal ownership structures based on economic interdependencies between the assets, $A = \{a_1, a_2, a_3, \ldots, a_n\}$ involved in the production of goods or services.

The formal deduction of these results is beyond the scope of this book, and can be referred to in Grossman and Hart (1996), Hart and Moore (1990). The following sub-section focuses on some selective results of the property rights theory of the firm that derive optimal ownership structures vis-à-vis economic properties of assets.

The results of the model

The results of optimal ownership structure for complementary assets, economically independent assets, and indispensable agents find interesting use when applied to explain the structure of the drug industry and bioprospecting.

Complementary assets (Hart and Moore, 1990: 1135)
Definition: Assets a_1 and a_2 are complementary, if they only have positive effects on the marginal productivity of the agents when they are jointly owned (ibid.: 1136).
Proposition 1
Complementary assets a_1 and a_2 should be jointly owned in order to maximize the incentives of specific investments by agents at date 0 which result in optimal production on date 1 (for formal proof, see Grossman and Hart, 1986; Hart and Moore, 1990; Hart, 1995).
Economically Independent Assets[21]
Definition: Assets a_1 and a_2 are economically independent, if their joint ownership has no influence on the marginal productivity of agents when compared with the marginal productivity of agents with separate ownership of these assets.
Proposition 2
Economically independent assets should be owned separately to maximize relation specific investments and joint production surplus. (for formal proof, see Grossman and Hart, 1986; Hart and Moore, 1990; Hart, 1995).
Indispensable Agents[22]
Definition: An agent i is indispensable to an asset a_1 if the marginal productivity of another agent j (who maybe owns different assets) is not improved by his ownership of a_1 unless agent i is the owner of a_1 as well (e.g. there is joint ownership of a_1 by agent j and i).
Proposition 3
An indispensable agent i should be the only owner of an asset a_1 in order to maximize her incentives for specific investment. Furthermore, by making the indispensable agent i the owner of the asset a_1, the incentives for ex ante investments are not reduced for the other agents j.

Ownership in the drug industry: Between vertical integration and contractual cooperation

Associated with the idea in property rights theory of the firm that complementary assets should be owned together is the idea that increasing returns to scale should lead to the formation of large firms. Hart (1995) states that the conventional concept of increasing returns to scale can be interpreted in the light of the property rights theory of the firm. Horizontal integration will occur if one large asset under single ownership is more productive than two assets half the size under separate ownership, because of complementarities between them (Hart, 1995: 51). Whereas this is the logic behind the series of mergers and acquisitions between major pharmaceutical firms in recent times, an interesting question stands out. The science-intensive nature of biotechnology no doubt led to the spawning of several smaller and medium-sized enterprises (SMEs) and dedicated biotechnology firms (DBFs) and inter-organizational collaborations between the pharmaceutical firms and these newer enterprises (See chapter 2). But what factors motivate the continued interactions between the large pharmaceutical firms and the SMEs and DBFs, given the transaction costs involved in external collaborations/contracts of this kind? Why do we not see total integration (Powell, 1996: 206)?

Powell (1990, 1996) studies the reasons for continued interorganizational collaborations in the drug industry. He identifies two major sources of transaction costs that are imposed by cumulative innovation – persuading the other party to make the appropriate investment in R&D and resolving the possibility of disputes over ownership of new products (Powell, 1996: 206).[21] Yet, interorganizational collaborations persist because the drug industry has developed operating structures that allow for organizations to rely on each other's competencies on the basis of a kind of mutual need (ibid.). This kind of mutual need at the macro-level is what Powell (ibid.: 209) terms as "virtual integration". He notes that:

> One way of accounting for the dense web of collaborative ventures is to see them as part of a strategy of accessing rather than creating resources or capabilities to complete the value chain from discovery to market place. Virtual integration, then, is a means for insuring long-term sustainability without incurring steep short-term risks (ibid.).

In the industry, a firm's reputation is tied directly to its R&D expertise and capabilities to cooperate meaningfully. Biotechnological firms are known for their overlapping, interdisciplinary project teams, with a high emphasis on scientific expertise.[22] Reputation and quality of cooperation are benchmarks for firms to prove their prowess, on the one hand, and to increase trust, on the other. The high probability of future association and increasing benefits of cooperation over time act as factors that minimize transaction costs that one would normally associate with such contracts.

Therefore, in these contracts, although the parties make relation-specific investments, disciplining opportunistic behaviour through reputation effects, coupled with competitive conditions for demand and supply, allow for efficient *ex ante* contracts to be written that clearly allocate ownership over results of discovery and extent of research efforts (Powell, 1996: 207; Holmstroem and Roberts, 1998: 85–86). High degrees of frequency and mutual dependency support cooperation across firm boundaries in biotechnology-based drug R&D.

Relation-specific investments on behalf of the firms in bioprospecting contracts

As noted earlier in Chapter 2, the two stages of increased demand for plant samples in the pharmaceutical research process are for preclinical and clinical testing and for large-scale production of the end product. Hence, guarantees of supply and quantity of supply of raw material become critical issues in the R&D process. Failing this, the entire R&D process, when dependent on the supply of raw material, can become susceptible to "hold-ups" after the initial stages of research, or, worse, after the development of the final product. More serious than ensuring the availability of such supplies during the research period is the issue of how to organize supply of the raw plant material after a product based on it has been discovered, in case this is needed for production of the drug. The firm always has to reckon on the possibility of finding the drug that may be dependent on raw material availability for its manufacture. Shifting partners during the R&D stages or even later is difficult for the firm because it will have to find "access" in other countries that have the same or similar species with sufficient concentration of active ingredients, and this is costly – or may not even be possible.

In these cases, both the factors that need to be fulfilled for a "hold-up" to occur are present in these contracts. Potentially, the entire R&D process is a relation-specific investment itself in case a drug is developed that relies on the availability of the natural product for its production, and, secondly, the exact amount of raw material that will be required to manufacture the drug or its quality, or even the time at which it is required, cannot be specified *ex ante* by the firms.

A good case that demonstrates this is Merck's experience with Brazil for sustainable supplies for Jaborandi. Laird (1993) notes that E. Merck of Germany had problems for over a decade to acquire samples of the substance pilocarpine from a domesticated shrub called Pilocarpus jaborandi which is found in North-East Brazil.[23] The case illustrates some of the difficulties associated with finding supplies of raw materials, especially when linked to sustainable harvesting and use. Pilocarpine is used in ophthalmology to reduce intra-ocular pressure, and is the treatment of choice for several forms of glaucoma. It was granted approval by the FDA, under the trade name Salagen, for use in the treatment of radiation-induced xerostomia. Merck had to discontinue the production of the drug for a period of about six years, due to problems in obtaining sustainable supplies from Brazil. The attempts made by the company for almost a decade to synthesize the plant were in vain, since laboratory synthesis proved to be a very expensive option when compared to harvesting the plants at their native place of origin (ibid.).

When the access country turns hostile, one option that the drug firms can explore is the cultivation of the same or a similar plant species in a different tropical country. For instance, in the case of the supply of Rosy Periwinkle for the production of vincristine and vinblastine, although the highest quality of the crude drug was harvested in Madagascar, the supply became unreliable due to problems between the drug company, Eli Lilly, and the native peoples, forcing the company to turn to other sources elsewhere for the supply of the raw material (see Chapter 6).

Even when we assume that the plant species in question can be cultivated in another tropical country, the irregular supply of raw material due to problems of negotiating long-term access with the alternate source country or the uncertainty relating to whether the crude extract will be as potent as required for manufacture of the drug may still prevail. Furthermore, at the early stages of R&D, where time as well as cost is a factor, the firm may not wish to take on the risk of investing in an independent supply of unexplored and unproven raw material. If Bristol-Myers Squibb had had to cultivate *Taxus brevifolia* in Texas in the first place, and wait until it was fully grown before conducting its clinical trials, the completion of the R&D process that produced taxol would hardly have been possible. An unusual initiative was taken up by the National Cancer Institute in the case of Michellamine B (see below), but it seems unlikely that private ventures would adopt similar initiatives.

The final intellectual property rights of firms are also contingent on legal certainty regarding the scope and content of intellectual property rights on traditional knowledge. That is, attempts to define traditional knowledge have to recognize that the knowledge can have different levels of contribution in modern inventions and clarify the status of inventions based on traditional knowledge. In the absence of this, firms always have to reckon with the possibility that they may spend years conducting research, developing a product, and applying for a patent, only to have a community seeking its revocation on the grounds that it has the

The case of Ancistrocladus korupensis and Michellamine B

Ancistrocladus korupensis is a woody climber generally found in the tropical forests of Cameroon and Nigeria. Early collections of this plant were made in the Korup National Park, Cameroon, in 1987 by staff of the Missouri Botanical Garden under contract from the Natural Products Branch of the US National Cancer Institute (NIC). In 1990, NCI research on the species showed promising activity against HIV in the *in vitro* process and, subsequently, an alkaloid, Michellamine B, that yielded the anti-HIV compound naphthyl-isoquinoline was isolated from the leaves of the species. Once development work on the compound reached the pre-clinical development stage, large quantities of the plant material were required, and this set in motion the search for the original species. After collections of *A. abbreviatus*, *A. ealaensis*, and *A. letestui* from central Africa failed to show activity against HIV, the search for the species continued until it was located by scientists of the Cameroon National Herbarium in 1991 in the Korup National Park. However, preliminary investigations by the NCI showed that not only was the plant not available in abundance within Cameroon, but also that harvesting of the plant in large quantities from within the Korup National Park was not allowed under Cameroon law. Additionally, harvesting trials conducted by NCI demonstrated that an interval of at least two years between leaf harvests would be necessary. In 1993, the NCI contracted Purdue University to work on the cultivation of *Ancistrocladus korupensis* at Korup, and this remains one of its largest investments in cultivation programmes to date. It was prompted by NCI's previous experience in respect of obtaining adequate supplies of taxol when work on the drug reached the clinical-trials stage. The NCI's letter of intent was also negotiated about the same time and signed by the University of Yaounde.

Subsequently, the research on Michellamine B had to be discontinued due to reasons of toxicity – the chemical showed neurotoxicity when tested in primates. But the case is a very good example of the many issues that can arise in bioprospecting in addition to those of uncertainty and the inherent risk in research based on genetic resources. For instance, it is an illustration of some of the legal issues that can occur in source countries. At the time of the *A. korupensis* negotiations the bioprospecting framework of Cameroon was uncertain; subsequently the government of Cameroon negated the Letter of Intent that had been signed by the University of Yaounde and the NCI *ex post* on the grounds that the university did not have the authority to sign such a document.

Additionally, the case illustrates some of the problems involved in managing intermediaries in collection processes. In this case, the increasing number of parties involved in the negotiations on plant collections created immense confusion in respect of their roles and responsibilities and, since NCI hired collectors to obtain leaf supplies, the specific details of benefit-sharing between the collectors and the local communities in respect of collections were not controlled directly by NCI.

Lastly, as Laird and Lisinge (1998) rightly point out, the case also shows how time gaps between original collections of the species and positive research results can be very great, sometimes spanning up to thirty years or more. This makes the legal frameworks for bioprospecting that protect the rights of source countries and communities over their genetic resources extremely critical.

Sources: Gordon Cragg (2004) personal communication; Laird and Lisinge (1998)

right over such knowledge or that the original contract struck between the firm and the access authority for the use of the said traditional knowledge did not have the community's consent.

Relation-specific investments of access authorities and communities

One could argue that the access authority has no specific investments in the drug R&D process because the access authority of the source country is just a holder of monopoly rights to access. The same could be supposed of the right to traditional knowledge. But in reality, if countries define rights over access to genetic resources and traditional knowledge in accordance with the CBD while catering to the challenges of harnessing their economic potential, the assets relevant for bioprospecting exhibit certain of the above defined economic properties of asset specificity. Why these assets are worth more within the bioprospecting contracts than outside is discussed in Chapter 6. This chapter will also use Propositions 1 to 3 of the property rights theory of the firm to deduce the vertical structure and contractual allocation of the control rights in the industry. It will also explore whether asset specific investments in bioprospecting lead to vertical integration or are better resolved through smaller contracts in competitive conditions as between the biotechnological firms and larger pharmaceutical firms.

VII. OWNERSHIP ISSUES AND INVESTMENT INCENTIVES IN BIOPROSPECTING

The market imperfections identified in this chapter can be because of two causative factors. One reason for the market imperfections is the absence of well-defined and enforceable property rights. To explore what property rights structures could be optimal for bioprospecting, Chapter 5 will focus on a law and economic analysis of the property rights over traditional knowledge, and Chapter 6 will similarly derive a well-contoured right for access.

The redefinition of property rights for access and traditional knowledge eliminates many of the market imperfections identified in this chapter. But those that relate to information asymmetries and uncertainty prevail. The second part of Chapter 6 deals with the design of optimal contractual mechanisms that can reduce contractual problems faced by parties in bioprospecting due to information asymmetries and uncertainty.

Notes

1. Other economic scholars like Jean Tirole define a complete contract similarly. According to Tirole (1988: 29): "A complete contract is a contract that has all the relevant decisions of transfer and trade incorporated within it, such as the expectations of parties, provisional mechanisms for dealing with all sorts of contingencies that arise during the contract, valuation of the good that is being traded and all sorts of costs associated with such trade."
2. In such cases, getting information about the nature of the dispute, the subject matter, and verifying the contingent claims may simply be too costly for courts to undertake (Hart, 1995).
3. For a detailed discussion of different kinds of transaction costs and their ramifications, see Furubotn and Richter (1998): 39–67.
4. The concepts explained in the next few sections can be found in greater detail in text books on microeconomics and information economics. The descriptions contained in this chapter are based on Mas-Colell (1995), Kreps (1995), Katz and Rosen (1998), Milgrom and Roberts (1996), and Scherer and Ross (1990).
5. See for example, Myerson (1991: 1–35); see also textbooks on microeconomics such as Mas-Colell et al. (1995) for a more extensive discussion on this topic.
6. To explain the concept of what the appropriable quasi value of an asset is, consider the following example. Subsequent to a favourable business offer by Y, a printing press owner X chooses to buy a particular kind of printing machine that is in demand mainly only by Y. According to the initial contract, X ought to be paid $4,000 daily, where in

X's operation costs are only $1,000. Imagine that X's investment would have a next-best use value of only $1,500 a day (the only offer for such printing machines by Z). This means that the minimum amount that Y has to offer to X to keep him running his printing machines for Y = the operating costs + the next best value = $1,000 + $1,500 = $2,500. This means that the appropriable value of X's asset is $4,000 – $2,500 = $1,500. Since Y knows that until and unless his offer for X's daily services falls below $2,500, X will have no choice but to work his machines for him; if Y chooses to behave opportunistically and breach the contract, he can try to appropriate the quasi-value of this asset, by altering the terms of the contract *ex post* and offering X an amount less than the initial $4,000 but more than $2,500. The problem that X faces in such a contract wherein his asset is relation-specific and whereby Y can behave opportunistically to appropriate the quasi value of this asset is what is referred to more commonly as the "hold-up" problem or the "lock-in" problem. For a "hold-up" problem to occur, two factors must be present – first, parties to a transaction must make relation-specific investment due to their long-term relationship, and, second, the exact form of optimal transaction (in terms of either the exact amount of units to be supplied, or the exact price for the units) cannot be specified *ex ante* due to uncertainty about certain states of the world that will only be known to parties *ex post*. Y is a publisher and he doesn't know when he contracts whether the book will be a flop or a best seller (due to uncertainty); he assumes the book is going to be a best seller, but, if not, then he has to reduce the costs.

7. For a good example of this sort of specific investment, see Klein, Crawford, and Alchian (1978).
8. These costs are cumulative – contingencies that may not have been provided for *ex ante* then add up to *ex post* costs since parties to the contract may have to renegotiate when the said contingency materializes in order to find appropriate solutions in the later stages of the contract.
9. As Williamson himself admits, "TCE maintains that each generic mode of governance – spot market, incomplete long term contract, firm, bureau, etc. – is defined by a syndrome of attributes to which distinctive strengths and weaknesses accrue" (2000: 606).
10. In economic literature, mechanisms that enable parties to overcome incentive constraints are called 'incentive-efficient' mechanisms. See Kreps (1995: 577) who notes, "Whenever there are informational problems of these or other sorts it is natural to ask, what is the best contract that can be devised?".
11. Moral hazard or hidden action refers to a situation in contractual relationships where a person acting in the capacity of an agent (be it a supplier, a customer, an employee) takes inefficient actions or provides distorted information (leading the principal to take distorted actions or waste resources) because the agent does not have the incentives to align his or her interest with the interest of the principal, and because the agent's behaviour cannot be monitored to the fullest extent by the principal.
12. Adverse selection refers to a situation where one party to the contract has much more information of relevance to the contract than the other. The problem is best characterized by Akerlof's market for lemons, where Akerlof uses the market for used cars as the example to explain the "lemons" phenomenon (see Akerlof, 1970). In economic literature, a classic example of a contract with adverse selection is an insurance contract where the insurance buyer clearly has much more information about the state of his or her health than the insurance seller. See Katz and Rosen (1998: 564) and Kreps (1995, Ch. 17).
13. The term "rent" in economic literature refers to any surplus. In this instance, informational rents simply means the additional surplus from the transaction that the party with the extra bit of information can extract.
14. Non-rivalry means that the usage of information by one does not diminish its avail-

ability and utility to another. Non-exclusiveness implies to the property of information whereby once it is made available to one person, it is also freely transmittable to others, or even that it becomes extremely hard to control its flow.

15. See Lisa Bernstein and her case study on the rules of trade in the diamond industry. This is similar to what has also been proposed by Willamson in his study of institutions. He refers to the possibility of parties devising their own "governance" structures when they transact with each other repeatedly since the benefits of such governance structures outweigh the costs of creating and maintaining the governance structures (see Williamson, 1979, 1985, 2000).
16. See also Klein and Leffler (1981) who analyse the role of repeat-purchase mechanisms as a measure of contract enforcement.
17. The property rights theory of the firm has its origins in Grossman and Hart (1986). It has been developed further by Hart and Moore (1990), Hart (1995), and Hart and Moore (1998). The entire GHM framework can be seen as a development of the earlier transaction costs literature of Williamson and others.
18. For an overview of the issues that govern the debate between Williamson and Grossmann–Hart–Moore, see Williamson (2000).
19. Starting from Coase's insightful exposition on why firms exist, this has been the major failing of attempts to explain the purpose of the firm. Some of the most notable of these theories are neoclassical theory, principal-agent theory, and transaction-cost economics. For a discussion of all these theories and their failings in explaining why firms exist, see Hart (1989).
20. This section is based on Grossman and Hart (1986).
21. Also see Rai (2001) and Gallini and Scotchmer (2002) in this context.
22. Powell notes that in a ranking of high-impact publications in the period between 1988 and 1992, out of the five institutions that were deemed to have been the best, two of them were biotechnological firms – Genetech and Chiron (cf. Powell, 1996: 200).
23. This paragraph is based on information contained in Laird (1993).

5

Intellectual property rights on traditional medicinal knowledge: A process-oriented perspective

Introduction[1]

The promise of intellectual property rights on traditional medicinal knowledge is recognition and benefit-sharing for communities in drug discovery and research, apart from incentive generation for biodiversity conservation. But control over resources is theoretical and incomplete if exchange of the resources is not possible or is somehow hindered. A central aspect of valuing any tradable resource is the right of action available on this resource and how these rights of action are to be enforced (Demsetz, 1964) – this is what links intellectual property rights and contracts inextricably in practice.[2]

From an economic perspective, contracts ideally have a two-fold effect: firstly, they should maximize the aggregate surplus that can be bargained upon, and secondly, they should allow for the distribution of this surplus in a fair and equitable way. But such welfare-maximizing contracts result only when specific conditions are satisfied. One precondition is the presence of well-defined, enforceable property rights that can be exchanged only by mutual consent of the property rights holders. Mutual consent ensures that the estimated value of the resource incorporates all the conditions of its usage (the opportunity cost of using the resource) apart from the subjective benefits that the user associates with the resource. (This includes social, cultural and spiritual values.) This condition for contract formation means that if communities associate high values with their resources, both tangible and intangible, they will not have to contract away such resources. The second condition for welfare maximizing contracts is the presence of equal bargaining power amongst the parties in the sense that there should be no inefficiencies resulting from imperfect competition (monopolistic or oligopolistic in nature) in the bargaining process. The third and final condition for the efficiency of private contracts is that of absence of externality effects, such as biodiversity depletion.

In the absence of these conditions being met, the right to traditional knowledge will suffer from problems of precise definition and demarcation of sets of beneficiaries, and will lack mechanisms that facilitate and mandate their free consent in the contractual process. This in turn would result in the value of the right – that is, any values associated by the communities not being fully incorporated into the economic price agreed as part of benefit-sharing.

The focus of this chapter is to find an operative definition of traditional knowledge for purposes of drug R&D. The chapter is structured as follows. Section I will enumerate the different issues that confront international and national decision-makers alike in the realm of traditional knowledge. Arguing for a sector-wise treatment for traditional knowledge, the section narrows down the focus to traditional medicinal knowledge only, and shows that the predominant issues that confront legislative efforts in protecting traditional medicinal knowledge – such as subject matter and identification of beneficiaries – are due to a lack of precise definition of traditional medicinal knowledge. Section II, through a discussion of the various theories in intellectual property jurisprudence, points out the convergence of both legal and economic justifications for grant of intellectual property rights, especially patents. Section III explores the contribution of traditional medicinal knowledge (defined as ethnobotanical knowledge) to the drug R&D process by highlighting the importance of tacit information within the process of cumulative innovation in biotechnology-based pharmaceutical research. The exercise here is to embed the contribution of ethnobotanical information within the theory of cumulative innovation, thereby making the case for allocative efficiency, as opposed to broader definitions that neglect such a perspective. By addressing the specific attributes of ethnobotanical knowledge and its role in drug research, Section IV discusses the need for a narrower, pinpointed definition of traditional medicinal knowledge components, based on the sector in which it is to be applied. The thrust is on being able to provide rights that are representative of the usefulness of traditional knowledge and that are capable of being traded, and on being able to create incentives through the grant of such a right.

I. TRADITIONAL KNOWLEDGE: TOWARDS CLARITY ON THE SUBJECT MATTER

The terms "traditional knowledge" or "indigenous knowledge" are used interchangeably by scholars in the literature and often in different senses (see below), based mainly on the academic disciplines and the different expectations placed on the right.

Definitions of traditional knowledge

The Convention on Biological Diversity (CBD) in Article 8(j) defines traditional knowledge as the knowledge, innovations, and practices of indigenous and local communities embodying traditional lifestyles relevant to the conservation and sustainable use of biological diversity.

Traditional knowledge can refer to, as the World Intellectual Property Organization (WIPO) uses it, "the tradition-based literary, artistic or scientific works; performances; inventions; scientific discoveries; designs; marks; names and symbols; undisclosed information; and all other tradition-based innovations and creations resulting from intellectual activity in the industrial, scientific or artistic fields" (WIPO 1998–1999: 13). Indigenous knowledge, on the other hand, is understood by the WIPO to be the traditional knowledge of indigenous peoples. Taken this way, "Indigenous knowledge is therefore part of the traditional knowledge category, but traditional knowledge is not necessarily indigenous" (ibid.: 23).

The words "indigenous knowledge" and "traditional knowledge" are used synonymously to "differentiate knowledge developed by a given community from the international knowledge system as generated through universities, government research centers and industry sectors" (Warren, 1998: 13).

Traditional knowledge or indigenous knowledge is also defined as the "unique traditional, local knowledge existing within and developed around the specific conditions of women and men to a particular geographic area" (Grenier, 1998: 1).

Traditional knowledge is also often used to denote indigenous medicinal knowledge that is "a coherent system linking social behavior, supernatural beings, human physiology, and botanical observations" (Reid, 1983).

All these definitions represent different facets of the universe of traditional knowledge, which are related yet distinct in nature, and are in keeping with the spirit of Article 8(j). But, as already discussed in Chapter 3, it is generally overlooked that Article 8(j) is a concept only and not an exercisable right in itself. Therefore, defining traditional knowledge within national contexts exactly or in a way similar to Article 8(j) does not help us derive a functional regime for traditional knowledge, due to its broad mandate. Such traditional knowledge could either be innovations or practices that are relevant from a commercial perspective, but even within this category, different forms of intellectual property-like protection may be necessary depending on whether it is medicinal knowledge or knowledge such as folklore and other forms of cultural expressions. Traditional knowledge and practices that relate to biodiversity conservation and management systems may demand a different mode of protection. In a similar way, the specific issues that confront the policy makers in relation to knowledge of farming communities are also different and need to be dealt with separately if a tailor-made regime is desired.

The sine qua non of successful legislative enforcement is to break up the mandate of Article 8(j) into successful compartments, and legislate vis-à-vis each one of them, according to the kind of knowledge in question. This is important for various reasons. To begin with, such a clear conceptualization is also the first step in resolving the interlinkages between the right to traditional knowledge and the rights over access regulation (Article 15 of the CBD) and patent rights of firms over biotechnological products as part of TRIPS Article 27(3)(b) (Straus, 1998: 119). Another reason why such a utility-based compartmentalization is extremely important is because the beneficiary categories may decidedly change from area to area, even apart from the property right that can be used to protect the said knowledge input. For example, the farming communities in cases such as the Basmati rice case[3] are decidedly different from indigenous communities whose medicinal knowledge needs to be protected through intellectual property rights. Identifying the subject matter of the right will automatically pave the way for decisions on, for example, identification of beneficiaries who will exercise the right, the link between the right and biodiversity conservation, or the kind of right that will best suit the demands of the knowledge category and the beneficiaries.

It is not the intent here to curb the scope of the rights of local and indigenous communities as recognized under various international instruments, from the International Labour Organization (ILO) Convention 169 to the CBD. It is a suggestion as to the means that should be used to achieve the various forms of recognition. For the larger goals of empowering indigenous and local communities, broader rights structures are required. But for the purpose of compensating their specific knowledge pools, whether relevant to drug research or some other field, more pointed rights and implementation structures need to be designed; failing which, broad and over-lapping property rights structures could hinder meaningful contractual exchange. This, in turn, frustrates the right-giving exercise.

II. JUSTIFYING AN INTELLECTUAL PROPERTY RIGHT OVER TRADITIONAL MEDICINAL KNOWLEDGE

According to utilitarianism, the prevailing school of thought in intellectual property jurisprudence, the nature of information and its usefulness (actual or potential) is the key criterion to decide who ought to hold and/or exercise the right over it. The function of a right is to grant appropriate incentives for the production and/or dissemination of the kind of information desired. Both these aspects translate into interesting propositions when applied to traditional knowledge. If there ought to be an intellectual property right on traditional knowledge, then this right has to be based on the potential contribution that it could have to the process of innovation in biotechnological drug R&D. This is the best clue to answering the question: which form of information are we are trying to protect? The answer will amount to a limitation on the right as well as on the set of beneficiaries who should exercise the right. Secondly, the "incentive" criteria of conventional intellectual property jurisprudence can provide us with a basis on which to differentiate between the different forms of information that constitute traditional knowledge, and which out of these ought to be protected, and what forms of institutional mechanisms are best suited for its production and dissemination.

The predominance of the utilitarian paradigm in intellectual property jurisprudence

Intellectual property is a tough notion to justify, even under normal instances, due to the elusive nature of information as a resource – when the use of information by one person does not affect or even reduce its availability to others, why should it be protected (Drahos, 1996)? If it is indeed protected, how and on what basis should a person be rewarded for the creation of information? To answer these questions, society has to sort out issues such as: does creation per se create a claim on ownership and, if it does so, to what extent is the creator the absolute owner of the information thus generated (e.g. May, 2000). If this is true and intellectual property rights are to confer absolute ownership rights over information, how and on what terms can such a right be overridden? It is here that the form of justification advanced for the intellectual property right assumes importance. Although almost all justifications of intellectual property rights are incomplete to defend such a right in its entirety, the kind of justification that is advanced is critical to determining whether such a right can be overridden or not.[4]

Contemporary explanations for the grant of rights in general, and intellectual property rights in particular, can broadly be classified into two main schools of thought – right-based theories and goal-based theories.[5] The right-based theories are also known as "natural rights" theories and the goal-based arguments commonly go under the name of "consequentialist arguments". Of the two, the goal-based theory/consequentialist argumentation is the more widely accepted basis for the grant of intellectual property rights. The juxtaposition of the individual's needs against the society's needs and making it the basis of granting rights, forms the core of goal-based theories for the grant of rights (Freedman, 1994: 380). The meaning of "goal-based" here is whether rights for something ought to be granted or not, and to whom they are to be granted, and this is decided on the basis of what needs to be furthered from a societal perspective (ibid.). Utilitarianism is the main paradigm on which such a goal-based/consequentialist argumentation is based.

Within the different kinds of intellectual property, the main view dominating patent law in particular is that since information is a public good, intellectual property rules and institutions are to be constructed to facilitate the production of specific kinds of knowledge within society (May, 2000: 27). Since production of information that can lead to societal progress and technological development has been the aim of almost all societies, intellectual property protection is justified on the trade-off between the static and dynamic gains of such information. As long as the gains to the society by encouraging the dynamic production of many types of information exceed its static costs of creation and distribution, it is desirable to design appropriate incentives for its production.[6]

Rai (2001: 5) notes that the law of intellectual property, for the most part, focuses on creating the right conditions for the production of information within the society to further its progress, and the economics of information analyses the conditions and costs under which information is produced and disseminated, which explains much of the common ground between the legal theory of intellectual property rights and the economics of information. A second look at information economics helps to clarify the unifying points.

According to Arrow, innovation is all about the production of information (see Arrow, 1962). Information economics posits that information as a good is characterized by non-rivalry and non-excludability. It is non-rival because one more person using the good would not affect the availability of the good to others, and it is non-excludable because the use of the information would result automatically in its divulgence (at least partially) to others. Therefore, knowledge, being a pure public good, would be under-provided for within the society, since the social returns to research investment exceeds the private returns recouped by the inventor. To rectify this under-investment in R&D (from the viewpoint of societal welfare) and to ensure optimal production of information, there is a need to internalize at least some of the social benefits. The grant of patent protection is meant to achieve this by creating a mechanism of exclusion, though imperfect, by giving incentives for individuals to invest in the production of knowledge in return for temporary monopolies (see generally, Scherer and Ross, 1990). This monopoly protection granted by patents creates a deliberate trade-off between the monopoly rents of the producers of the knowledge and the consumer welfare of the society at large.[7]

Arrow's contribution was seminal to the theoretical economic framework on information and prompted a wave of economics literature arguing for the strengthening of intellectual property rights in order to promote the production of information in society (Winter, 1989: 41). Arrow's analysis focused on the elusiveness of information due to its non-rivalrous and non-excludable nature and stressed the need for an incentive scheme to ensure its optimal production in society.[8] According to Arrow, the continuous paradox for the society remains one of creating institutions that rectify the disproportionate division of private versus social gains, so that production of information is encouraged. Institutions of intellectual property, patents in particular, walk this thin line of balancing private rewards with social benefits. Hence they have been the subject of much tension and debate, academic, policy-related, and otherwise, even within the neoclassical perspective.

The right-based theories versus the utilitarian paradigm

In contrast to the utilitarian justification, right-based claims for intellectual property in general are based upon arguments of labour or the instrumentalist justification,[9] individual self-assertion,[10] self-development,[11] the idea of sovereignty,[12] or even moral individualism.[13] Of these, the strongest are discussed here.

The *instrumentalist justification* is based on John Locke's labour theory that people are entitled to the fruits of their own labour (Hettinger, 1999:

122). Because it assumes that individuals create 99 per cent of the value of the resource, it is believed that Locke's justification for property rights is probably more applicable to intellectual property than any kind of physical property (ibid.: 123).

Another popular natural right-based argument is the self-development justification, based on Hegel's idea. According to this argument, an individual's free existence and respect within society is based on the legitimacy of his or her property claims, therefore, the recognition of property rights is an integral part of the freedom of individuals "since the respect others show of his property by not trespassing on it reflects their acceptance of him as a person" (May, 2000: 26).

The inadequacy of right-based justifications

Despite the fact that natural right-based justifications are based on a variety of primal concepts, they are limited in explaining the nature, scope, and extent of intellectual property institutions in today's society. Even the Lockean approach that stresses the need to reward creators for their creations through ownership proceeds when intellectual goods are exchanged, which is probably the most accepted right-based justification, has its own limitations when arguing in favour of intellectual property. To begin with, as Hettinger notes, "Even if one could separate out the laborer's own contribution and determine its market value, it is still not clear that the laborer's right to the fruits of her labor naturally entitles her to receive this" (Hettinger, 1999: 38). Locke's justification is also problematic because the author and his creation are granted so much sanctity in the property rights' exercise. Lastly, while such a justification can be applied to certain forms of intellectual property like copyrights, it cannot be made applicable to other forms of intellectual property, such as patents (Drahos, 1996). This inadequacy is evident by the amount of confusion there was within patent law right from the start. Drahos (1996: 29) notes the confusion in English law about the nature of the act for which a "patent" should be granted – authors create, as a result of which they are to have a right to property on their creation, whereas inventors get a "patent" because they only uncover what is already there.

A second but equally grave issue in granting rights completely based upon moral justifications is the question of conflict of rights. Since the possibility of overriding rights depends on the basis on which they have been defended, if rights are stated as moral claims and presented as pre-eminent, it is difficult to determine whether a right can be overridden at all, and if so under what situation (Freedman, 1994: 381). As Freedman (1994: 381) asks: "What happens when rights conflict? Are there any situations and if so, which, in which such rights may be 'trumped'?"

Intellectual property rights over traditional knowledge: The justifications

Up until now, the major defence for intellectual property protection for traditional knowledge has come from a natural rights-based perspective. Intellectual property rights for indigenous peoples have been justified on the basis of a system of entitlement theory[14] and theories of self-development, and as value of individual autonomy.[15] Based on natural rights justifications, several systems of protection have been proposed for traditional knowledge. This includes a system of traditional resource rights (Posey, 1996), a system of discoverer's rights (Gollin, 1993), a system of identification of source materials (Gadgil and Devasia, 1995), and a system that advocated separation of ownership of genetic resources from the ownership of the knowledge itself (Lesser, 1996), amongst many others.

The lack of an information economics perspective and its repercussions on traditional knowledge

The two general drawbacks of moral justifications are all the more aggravating in the case of traditional knowledge. Firstly, moral justifications do not provide guiding parameters for either demarcating the resource that should be protected and made transaction-worthy or the set of beneficiaries who should be entitled to share benefits by way of the right. This drawback hinders meaningful exchange of the intellectual property right – a precondition for benefit-sharing to occur.[16] The issue of not being able to reconcile the right with other rights in times of conflict of rights is also serious, especially because, when viewed within the process of drug R&D, the right to traditional knowledge has to be functional and flexible vis-à-vis its interactions with the other rights in question. Failing this, issues of overlap between the right to traditional knowledge and other rights can, in the process, hamper efficient contract formation.

This does not mean that moral justifications cannot be a basis for the grant of an intellectual property right over traditional knowledge,[17] but that it cannot be the sole basis on which the right is based and defended.

The second major drawback in the attempt to derive a right based on moral considerations is that it fails the set criteria of information economics. Since traditional knowledge is knowledge that has already been produced, one could argue that there is no need to protect this right through intellectual property protection. This is because knowledge or information, once produced, is a public good. The diffusion of the information among the members of society can be achieved at negligible marginal cost and thus the optimal equilibrium price for such information should be close to zero. A simplistic economic analysis could argue that already produced traditional knowledge is such an information pool, which could and should be made available freely to potential users like researchers or firms. This economic analysis would prevail if the knowledge could be provided at negligible cost.[18] Needless to say, traditional knowledge would pass the test of efficiency only when the answer to this question is in the negative.

The ineffectiveness of the justifications for traditional knowledge makes the best argument to reassess the status of traditional knowledge as an intellectual property right. It would not be wrong to argue that in order that traditional knowledge is granted intellectual property right, two important criteria have to be satisfied. Firstly, the right has to pass the test of incentives – that is, what are the incentive effects of granting the right to traditional knowledge? And more importantly, it has to be that the dissemination of this already existing knowledge should not occur at negligible cost.

It is not so much the argument here that there has to be the conventional incentive to "produce information" when an intellectual property right is sought, but that there has to be some alternate incentive effect that such a right ought to achieve – otherwise the argument for the grant of an intellectual property right for information would be vitiated.[19] Such a definition would make it easier for us to determine the scope and nature of protection, thereby paving the way for the design of appropriate institutional mechanisms for its exercise.[20] The property right itself has to be one of the key criteria in the identification of the beneficiaries, as is the case amongst other forms of conventional intellectual property categories. For this, it is critically important to determine the contribution that an intellectual property right to traditional knowledge can have towards (a) drug research and development, and (b) biodiversity conservation, and to base the right upon it.[21]

III. THE PROCESS OF INNOVATION IN DRUG RESEARCH AND PLAUSIBLE INCENTIVE EFFECTS FOR TRADITIONAL KNOWLEDGE

The application of biotechnology as a research tool across various industrial sectors has had far-reaching implications for the value of technological information within the R&D process.[22] The changed nature of innovative activity induced thereby has major implications towards finding a more convincing argument for traditional medicinal knowledge. The industrial structures in these fields of technology are characterized by the important role that availability of knowledge inputs and their transfer, in either tacit or codified forms, play in further production of knowledge.

Therefore, if it can be asserted that traditional medicinal knowledge qualifies to be what can be called "tacit" inputs to this sequential innovation process, it provides a very convincing basis on which to argue for an intellectual property right on such knowledge. To find out whether this is the case, this section begins by assessing the changed nature of innovative activity and the importance of tacit information in the process of cumulative innovation for drug research. A part of this inquiry is directed towards economic theory of innovation and the critiques on the inadequacy of the neoclassical theory to cope with the functional attributes of information in the newer forms of industrial organization. The focus then is on parts of traditional knowledge – those parts that are specifically called ethnobotanical knowledge, which can be called "tacit" inputs in this process.

The process of cumulative innovation in drug R&D

Innovative cumulativeness, according to Scotchmer (1999: 1) means "innovators build on each other's discoveries". Cumulative innovation therefore refers to a pattern of innovation wherein the innovation at the end of each stage is an essential input for reaching the next stage.

Of the many recent technologies that demonstrate this sort of innovative cumulativeness, biotechnological research is a prominent one – as already discussed in Chapters 2 and 4.[23] Biotechnological innovations are cumulative in nature, characterized by the incremental value that the product gains as it moves up the R&D chain, "most of the sequential innovation occurring in the pre-commercial stage, before the initial marketable product, most often a drug, is produced" (Scotchmer, 1999: 1). This nature of biotechnological research has changed the focus of knowledge enterprise – from sheer knowledge creation to knowledge access and transfer as well.

Patent policy and its emphasis: An evolving concept?

There are two critical features of cumulative innovation that do not figure in a dominant way in the traditional literature in economics of information:

(a) the relevance of already produced knowledge, largely tacit knowledge, for the production of codified knowledge, where codified knowledge refers to knowledge of the kind that can be protected through modern intellectual property instruments as we know them (Mandeville, 1999: 46);
(b) the cumulative nature of knowledge – that is, every advance being strongly dependent on the accumulated stock of knowledge and the evolution of "knowledge streams" (Foray, 1995: 84).[24]

These emphasize the large importance within biotechnology-based sectors, not only for incremental knowledge creation, but also for the transfer of these complementary assets through what are called "knowledge-intermediaries".[25]

This changing reality of innovative activity has set into motion a criticism of the conventional approach to information. Starting with Nelson (1959) and Arrow (1962), other neoclassical scholars who espoused the theory of innovation tended to look at research as an investment decision of the profit-maximizing firm and the impetus for this investment decision lay in expected returns. This inadvertently led to the assumption in patent literature that innovation denotes unconnected events, even within the same field of technology, and that the disclosure requirement in patent law was sufficient to tap the indirect impact that any given invention could have on future innovations. Mainly due to this, economic theory of intellectual property rights has been criticized to "limit its applicability to modern controversies in the field of biotechnology, computer software and computer hardware" (Scotchmer, 1999: 1).

Not only does the description of information as a "public good" not take into account the knowledge-intensive nature of exploiting already produced information, neoclassical treatment of information also undermines the critical role that the utilization of the results of some other R&D activity and associated information, whether in tacit or highly codified forms, played for commercially successful innovations (Rosenberg, 1982: 6).[26] Zeckhauser (1996) observes that the public good description (of non-rivalry and non-exclusiveness) might be an adequate description for some types of information, particularly comparative in nature – such as the scores of sportive events. But he claims that this is certainly not true if we are dealing with technological information. He notes that "the focusing on the public good aspect of information has deterred economists and policy analysts from delving more deeply into the distinctive properties of information, including most particularly, the challenge of contracting for technological information" (Zeckhauser, 1996: 12743).

In the context of cumulative innovation, the neoclassical construct that technological information as something that is created mainly within the isolated firm is not only considered to be unrealistic (Mandeville, 1999), it is also considered to run contrary to the reality in modern technological processes wherein transferring and exploiting information required for innovative activity is by itself a costly and intensive process (Rosenberg, 1982: 6). Based on this, scholars have argued often that the costs of acquiring information and the information properties of markets and organizational structures that condition the exchange of information are central issues to resolve while designing intellectual property mechanisms.

The underpinnings of modern innovative activities

As Gallini and Scotchmer (2002: 53) note, "Arrow explained *why* some incentive scheme is needed, but not *which* scheme". There should be enough room to identify the changing nature of innovative activity and to devise tailor-made incentive schemes that capture its needs to the fullest. There are newer demands on policy-making due to the cumulativeness of innovation; demands that stem from the extensive reliance on building on already existing knowledge inputs.[27] The dilemma in intellectual property policy making in the traditional model was to find ways to reconcile the two private contradictory demands of protecting the inventor's needs and the society's need of knowledge dissemination. In contrast, modern innovative activity lends strength to the proposition that the generation of new knowledge and the distribution of such knowledge are not contrasting policy aims; the performance of the system would only improve if access to existing knowledge stocks would function in a more efficient way (Foray, 1995: 89).

Such efficient access to and dissemination of already created knowledge stocks consists of two parts – the grant of incentives to agents who hold useful information to reveal this information and the creation of efficient mechanisms for the trade of such information. The repercussions of the lack of legal and contractual mechanisms for easy and cost-effective access to such knowledge and its trade are not to be underestimated. In the absence of adequate policy initiatives that allow the former, even when there is no direct restriction on access to information, it may be extremely costly to acquire (Zeckhauser, 1996: 12475–12476). Disputes relating to innovation may become commonplace when there are non-existent or ill-functioning mechanisms for trading information.

All this is more than clear when one takes a closer look at biotechnology-based R&D. As discussed earlier in Chapter 2, biotechnology-based drug research is characteristic of a new form of industrial organization that is science intensive and exhibits a high reliance on inter-institutional collaborative agreements (Saviotti, 1998: 19). Agreements and ventures that are popular in this field of technology clearly show that, first, these new technology paradigms are more knowledge intensive than before, and, second, successful innovation relies on the capability to acquire more information on what is going on in the field (Archiburgi and Michie, 1995: 128–129). This acquisition of information on what is relevant for the field largely entails identifying and contracting for both tacit and codified information inputs that can contribute to the R&D process.

Quantifying the importance of tacit information in general and traditional knowledge in particular

Given that acquiring and building on knowledge incrementally in cumulative innovation is a process wherein information is gathered from a variety of external sources (Mandeville, 1999: 44), one is tempted to inquire into the nature of such information. What sort of information are these bits and pieces that have an incremental effect on the building of such knowledge trajectories?

Innovations as improvements of earlier products: The relevance of uncodified information

It has been proposed that in the context of cumulative innovation, technological information be viewed along a continuum wherein the nature of information ranges from highly uncodified to highly codified. According to Mandeville (1999: 46), codified information on this continuum refers to information that has been recorded and put into action – for instance, highly uncodified information becomes codified when products, such as machines, have been derived out of it, and the relevant information has then been formalized through blueprints or patents, etc. This is why he refers to such information as "tangible" although it still refers to intellectual property.

On the contrary, highly uncodified information refers to intangible information that may include either undeveloped ideas or particular know-how of any sort, which is best communicated through personal communication between people (Mandeville, 1999: 46). Making the distinction between the two forms of information would necessitate the realization that the public good nature of information, as made out by neoclassical theory, would more likely refer to information at the far right of the continuum (which consists of the highly codified category), wherein the information that has been accumulated, produced, and recorded, and therefore can be replicated at negligible cost (ibid.: 47). In contrast to this, tacit information entails high costs of acquiring information and exploiting it for newer technologies that may benefit from it.

As in any other technology, tacit information plays a critical role in biotechnology-based drug research too. Recalling from Chapter 2, interorganizational collaborations between the large firms and the SMEs and DBFs are highly relevant in this sector. Within these organizational structures, the functional roles for the SMEs and the DBFs in the drug industry can be one of: an independent innovator, a niche operator (to exploit a very specialized niche), or a problem-solver using biotechnology methods for specific areas.

In their endeavour to be competitive suppliers of knowledge services in these three fields, SMEs and DBFs rely on acquiring and processing large amounts of tacit information – which can be in the form of technological skills or the usage of plants/animals/microorganisms among other genetic substances, and so on.

Traditional knowledge as a "complementary asset" in cumulative drug R&D

Capturing the value of ethnobotanical knowledge to modern drug research lies in being able to harness the complementarity between traditional and modern knowledge for drug R&D, and more importantly, in defining property rights so as to best address this complementarity (Rausser, 2001, personal communication).

Ethnobotanical technique is one targeted search technique that firms could rely upon to narrow down upon genetic resources for R&D programmes. Viewed within the process of cumulative innovation, ethnobotanical knowledge fits into what is called tacit, non-codified information that has a specific contribution to make to the R&D process. In order to transform a plant into a medicine, informational details such as the correct species, its location, the proper time of collection (some plants are poisonous in certain seasons), the solvent to use (cold, warm, or boiling water, alcohol, etc.), the mode of preparation (time and conditions to be left), and finally, the posology (route of administration, the dosage) are extremely important.[28] Such ethnobotanical knowledge is tacit knowledge because it is knowledge that is "acquired experimentally and transferred by demonstration, rather than being reduced immediately or even eventually to conscious and codified methods and procedures".[29] Furthermore, it entails more than the transfer of just one element; it is the transfer of many elements of the system, making it fit better into the "tacit information" category (see for example, Winter, 1987). According to Cox (1995: 10), this pattern of events consists of five stages: "specialist indigenous healing knowledge is recorded by interviewing indigenous healers; the plant materials used by indigenous healers are collected and identified; these plant materials are screened for pharmacological activity; the molecular entities responsible for the observed pharmacological activity are isolated using bioassay-guided fractionation; and lastly, the purified materials are structurally determined using extant technology."

As mentioned in Chapter 2, viewed within the process of cumulative innovation for drug R&D, ethnobotanical knowledge can be a very useful input with two distinct contributions. This knowledge comes into question before the screening of compounds starts within drug research and can have two distinct contributions to drug R&D programmes. For one, it can make the research process cost-effective by providing easy and efficacious means of short-listing plants or "pre-screening" that could be tested for higher pharmacological activity (Aylward, 1995: 115). Secondly, traditional knowledge itself could be the end use on which the product is based – this is a possibility when one talks of botanical medicines for common ailments such as the common cold, or fatigue, etc., or other health supplements.

Estimates of ethnobotanical contribution to modern drug research

Although proof that ethnobotanical knowledge has much to contribute to modern drug R&D has been advanced by many writings in law, anthropology and ethnobotany, the evidence gathered is unsystematic and is often based on historical anecdotes rather than any futuristic predictions. One of the more specific estimates of the contribution of ethnobotanical knowledge to drug research is that by Ried et al. (1996: 150), who note that using ethnobotanical knowledge actually increases the probability of drug discovery fourfold.

What makes this estimate more appealing is that many other studies that use economic and statistics tools have proposed the same probability estimate, albeit representative of a very small proportion of all the research in this field.[30] One interesting study that arrives at a similar conclusion is Balick's study at the Institute of Economic Botany of the New York Botanical Gardens for the National Cancer Institute, USA, to collect 1,500 plant samples for its anti-cancer and anti-AIDS screening programmes. Most of his work for the National Cancer Institute consisted of collecting ethnobotanical samples in Belize. Based on his experience on random versus ethnobotanical collections of samples in this project, Balick conducts an Ethno-Directed Sampling Hypothesis by comparing two groups of plants that were collected and sent to the National Cancer Institute for screening against HIV-AIDS. The random collection included plants from Honduras and Belize, although Balick admits that the results are preliminary and that only data from a small number of samples are available for analysis, the data which has been analysed using the Fisher Exact Probability Test (a methodology useful for handling data from small-sized samples) shows a fourfold increase in probability when ethnobotanical information is used in such research instead of random searches.[31] The results of this test show that $P = 0.101$. Although this is not conclusive evidence, it is a strong indication that the null hypothesis (that there is no relation between ethnobotanical collections and active compounds) should be neglected (Balick, 1990: 26–28).

Apart from this, there are other studies that report the use of such knowledge in discovering novel compounds and can confirm it up until the stages of clinical trials (for products pending these trials).[32] But estimates are rather the exceptions in ethnobotanical studies. Ethnobotany, which could perform the task of quantifying the contributory capabilities of such knowledge falls short of doing so because of its methodological constraints. Most of the ethnobotanical studies do evaluate the plants used by indigenous groups in detail and list their uses, but do not validate them through laboratory testing, or any other sort of modern medical methods. There is perhaps a need for more interdisciplinary surveys in this field, which combine ethnobotanical findings with statistical tools.

Traditional knowledge as "tacit information" in the cumulative innovation process

Apart from the limited evidence presented above, the temptation to simplify traditional knowledge as an issue for distribution is also negated by the behaviour of the drug industry, which itself seeks out sources of such information from journals or similar sources.[33]

Ethnobotanical knowledge is not readily available for dispersal at marginal cost. The information that is not already documented in journals and other publications, which is the case with over 90 per cent of this kind of knowledge, requires collaborative activity – either by scientists, ethnobotanists, or even national agencies – that can document such knowledge and map it on a one-to-one basis to modern medical science. This being the case, within the scheme of complementary assets in cumulative innovative activity (Winter, 1987), ethnobotanical knowledge qualifies to be called a "complementary asset" for the innovation process in drug research based on natural products.

This role of ethnobotanical knowledge within the knowledge chain of drug R&D, as tacit uncodified information within this process, gives us one of the most powerful arguments to defend such an intellectual property right. Industry counterparts presently seeking ethnobotanical information rely primarily on published information, often available through databases such as NAPRALERT and MEDLINE (Laird et al., 2002: 93). It is not clear, however, as to how effectively ethnobotanical knowledge can be transmitted through databases and most other publications due to the complexities of these systems and difficulties in fully documenting them (see Chapter 6). In principle, however, researchers continue to hold an interest in ethnobotanical knowledge associated with species of particular interest, and will seek information on traditional uses through databases and publications, and some collecting programmes use ethnobotanical knowledge as a tool to aid collections (Sarah Laird, 2004, personal communication). The lack of a thriving industry in this "niche service" is mostly attributable to international and national policy issues and lack of mutual trust amongst contracting partners rather than anything else.

IV. AN INTELLECTUAL PROPERTY RIGHT OVER ETHNOBOTANICAL KNOWLEDGE: THE INCENTIVE EFFECTS

Such an intellectual property right can be defended on the basis of two incentive effects – the incentive effect to reveal the knowledge, thereby reducing the cost of acquiring it, and the incentive effect to keep the knowledge pool in its entirety. Such an intellectual property right can also act as an incentive to conserve biological diversity, when coupled with recognition of autonomy on the tangible land occupied by communities.

Incentive effect 1

It has repeatedly been noted that, as a result of researchers and/or private individuals appropriating the knowledge for purposes that are not agreeable to the communities or, in many cases, not even made known to them, many of the communities fear the misappropriation of their information against their economic, social, and cultural values. Grant of such a right (when supported by appropriate institutional mechanisms) ensures that they are, in fact, in a position to control the usage of the knowledge, thereby giving them an incentive to reveal useful knowledge.

Incentive effect 2

Communities often maintain their own indigenous forms of intellectual property right structures, wherein common remedies and usage of plants are known to all members but the more complex systemic parts are held in secret amongst a select few (Huft, 1995: 1697; Daly and Limbach, 1996). In the absence of a community right over such knowledge there may be incentives for individual parties to divulge parts of the knowledge known to them, thereby endangering, on the one hand, the holistic nature of these knowledge pools passed on from one generation to another, and, on the other hand, the structure of the community itself.[34] Therefore, a second and equally important incentive of such a right would be that when a community intellectual property right is granted, it mandates that the communities and its constituent members devise mechanisms to enforce it as a whole, thereby ensuring that the knowledge pools are kept intact and community structures are fostered to the extent possible.

The right vis-à-vis the biodiversity conservation

The main stimulus for the recognition of rights of local and indigenous communities in Article 8(j) is based on biological conservation. There is ample evidence that such communities serve as guardians of their immediate environment most times, and that they also possess specific knowledge of conservation and sustainable use mechanisms that can be of much use to us. Proof of contributions of traditional knowledge to sustainable use and conservation is abundant – ranging from systems of indigenous management (Mundy, 1993), environmental impact assessments (Sallenave, 1994), seed selection and preservation (Bandhopadhyay and Saha, 1998), and breeding of animals (Wu, 1998), to other methods of hunting, fishing, or agriculture (Warren, 1991).

But, at the same time, as Kothari and Das (2000: 190–194) note, communities tend to overuse. Alternatively, in the tough situations that local and indigenous communities live in today, the realization that their knowledge about such resources is profitable may actually lead them to be co-accomplices in exploiting these resources.

Despite this, local and indigenous communities demonstrate common property resource management in territories that they occupy. Common property resources have sustained such local and indigenous communities either by directly providing resources that can be exchanged through market mechanisms, or by just enabling them to sustain themselves on the basis of such resources. The dependence of the communities on the resources leads to internalization of social sanctions, thereby creating conditions for regeneration.

The hypothesis of welfare maximizing norms suggests that common property management is based on the ability of private agents to cooperate in order to invest together in sustainable and prolonged use of the resource such that each of them could benefit better for a longer time from resource use.[35] Usually, in common property management, community mechanisms exist that control the extent to which the resources are used together with sanctions on deviance. The ability of the members of the community to exclude the persons (mainly outsiders but, in situations of misconduct, even the members of the group) from using the resource is the main characteristic that is decisive in terming a resource a "common property resource".[36] This facilitates the creation of internal rules and control mechanisms over the resources occupied by the group.

Accordingly, if communities which possess traditional medicinal knowledge/ethnobotanical knowledge are vested with the right to physically control their territories, as argued for in the previous sections, it would help them to impose limitations for the usage of biological resources within their territories, which is the crux of achieving such community-based conservation. Empirical studies also widely recognize and accept that access restrictions have been the key instrument in earlier times that have helped such communities manage their resources.[37] To this extent, the proposal that traditional knowledge should be defined much more narrowly, and in keeping with the aspect of autonomy that such communities require to exclude outsiders from the use of their resources, would also serve best the goal of biodiversity conservation.[38] The UZACHI (discussed in Chapter 2) are a good example of a community that has a sustainable management plan that covers both their commercial and conservation activities and which is managing them successfully through common property resource mechanisms (Chapela, 2002, personal comunication).

Other attributes of the sui generis *right*

The major contribution of this chapter has been to show that successful attempts to define traditional knowledge ought to focus on demarcating the nature of the contribution that such knowledge could have rather than on the physical attributes of the right itself. The emphasis on the nature of the information itself serves as the best parameter of what the limits of "community/communities" are and what sort of knowledge ought to be protected and made contractible through an intellectual property right. The chapter has undertaken this exercise for traditional medicinal knowledge and shown that allocative efficiency issues for granting of such a right do, in fact, exist.

After thus asserting the nature of the contribution, countries can choose from amongst a wide range of intellectual property instruments. The most effective options to protect traditional medicinal knowledge/ethnobotanical knowledge seems to be the options of trade secret protection or a system of *sui generis* community intellectual property rights.[39] Categories of traditional knowledge that do not fall within the abovementioned criteria could be documented in traditional knowledge databases to prevent third parties from claiming patents on already existing knowledge. Making such public domain available as "prior art" will ensure that third parties do not unduly benefit by claiming patents on such information.

The trade secrets option seems to be especially suitable for the protection of ethnobotanical knowledge for two reasons. Firstly, it affords protection for informational goods that do not fit into the strict standards set out by patent law. Secondly, Article 39 of the TRIPS Agreement has made trade secrets an internationally acceptable intellectual property option.

Before the adoption of the TRIPS Agreement, countries had their own ways of dealing with trade secrets, a majority of them treating infringement of a trade secret only as a harm to the holder of the secret, thereby prescribing damages or injunctions as legal remedies (Verma, 1998: 729). Now, as per Article 39(2), exploitation of a secret without the consent of the owner is improper (contrary to honest practices). But since "improper" in this context depends on breach of contractual restrictions, a large onus in making the protection work rests on effective contractual restrictions (Krasser, 1996: 223).

Another advantage of Article 39 is that the member states are free to adopt effective measures of protection under national laws. Thus, if groups of persons who shared ethnobotanical knowledge could be organized as legal persons, and if "reasonable steps ... by the person in control of the information, to keep it a secret" (Article 39(2)(c)) can be proved, ethnobotanical knowledge can be protected as a trade secret. It would also make sense for contractible ease, since much information in the biotechnological industry is kept as a trade secret.

Article 39's weaknesses lie in its lack of protection against third parties and accidentally disclosed secrets. The Article does not afford any protection to holders of trade secrets when persons who are not in contractual relations with the lawful holder/user of the information acquire information through lawful means. But, where third parties previously knew about the confidential nature of the information or were grossly negligent in failing to know that dishonest commercial practices were used to acquire the information, they may be held liable (Verma, 1998: 730). Since the definition of the terms "gross negligence" or "knowledge" is left to national jurisdictions, this discretion could be used to protect communities against dishonest practices by governments or private agencies who document their knowledge as trade secrets (or even their own compatriots who attempt to sell their knowledge as trade secrets). The lack of protection against accidentally disclosed secrets could be a considerable issue when applied to traditional medicinal knowledge given that such accidents may be hard to prevent in communities. A framework for implementing the trade secret system for traditional medicinal knowledge at national levels is discussed in Annex 1.

But a well-defined right with an easily segregable set of beneficiaries is only the first step in empowering local and indigenous communities in this regard. A large onus rests on the design of institutions that will put this right into an enforceable framework. Whatever be the mode of intellectual property option chosen for the right (like a trade secret or a know-how licence), the institutions have two major tasks: that of representing the communities effectively, and that of providing for rules of contract formation which take into account the difficulties of dealing with information as a resource.

The task of representing the communities effectively, in turn, has two main components. First, the institutional framework has to provide mechanisms to minimize principal agent problems that can arise between the communities and their agent, the state. The second and equally important aspect is that of ensuring the unified participation of the communities and their members in decision-making processes. There are many shortcomings of a community right, which the institutions have to take into account in order to facilitate this. One shortcoming relates to the gradual breakdown of control mechanisms within community structures in transitory phases due to modern inroads or heterogeneous elements making resources more prone to unsustainable exploitation. Another potential shortcoming that could hamper unified decision-making is caste–creed conflicts within communities – for example, as is the case amongst communities in India. It could also be that knowledge holdings are segregated within communities, making the people who possess the relevant knowledge more important than the rest. As a result, the successful implementation of the right for the welfare of the communities and societal welfare rests on the capability of the institutional mechanisms to anticipate and prevent such issues of enforcement.

The second task of the institutional mechanism is to provide contractual rules that can take into account problems caused by information asymmetries, uncertainty, and transaction cost issues that parties face in evaluating and trading with traditional knowledge. These issues that confront national policy makers in making the right enforceable are discussed further in Chapter 6.

Notes

1. This chapter was first published as an article in the *Journal of World Intellectual Property* 7(5), September 2004.
2. See Nimmer (1998) for an extensive assessment of the relationship between intellectual property rights and contracts in the case of software.
3. "Basmati", an aromatic variety of rice grains that is traditionally grown in hilly regions of Northern India and Pakistan along the foothills of the Himalayas, was the subject of a US patent. The patent on this variety that had been granted to the firm Ricetec has now been revoked (Dutfield, 2000: 88). See also, Lightbourne (2003) for a detailed analysis of the issues raised by the Basmati case.
4. See May (2000) generally for a discussion of different forms of intellectual property rights' justifications. See Freedman (1994: 383) who notes, "We cannot determine whether a right may be overridden until we see how it is justified. We should look not to its formulation, but to the values which underlie it."
5. See Freedman (1994: 379–392) again for a detailed discussion of rights' jurisprudence.
6. See also Calandrillo (1998) for an analysis of various justifications of intellectual property rights in the US context.

7. In other words, the monopoly protection for a limited time would imply that the producer would be in an advantageous position with respect to his product and there would be a considerable loss to the consumer group by way of a monopoly price, but the system intends that within the period of a monopoly, the inventor's sunk costs of investment need to be recuperated, and the consumers can derive complete benefit of the product upon the expiry of the monopoly.
8. According to Arrow: "no amount of legal protection can make a thoroughly appropriable commodity of something so tangible as information. The very use of the information in any way is bound to reveal it, at least in part. The demand for information also has uncomfortable properties. In the first place, the use of information in certainly subject to indivisibilities ... In the second place, there is a fundamental paradox in the determination of demand for information; its value for the purchaser is not known until he has the information, but then he has, in effect acquired it without cost" (see Arrow, 1962: 615).
9. Based on John Locke's Treatises.
10. See H. L. A. Hart (1955) discussed extensively in Freedman (1994): 388.
11. The self-development justification draws on the work of Hegel, as spelt out in the *Philosophy of Rights* (1967: 1821) and elsewhere, discussed extensively in May (2000): 26–28.
12. See the discussion on Austin's notion of sovereignty, in Freedman (1994): 216–221.
13. This is one of the intellectual underpinnings of the law of contract, more popularly known as the will theory (see Freedman 1994: 388).
14. As Stenson and Gray note, this view holds that communities are "morally entitled to intellectual property rights in plant genetic resources and associated knowledge." For a detailed discussion, see Stenson and Gray (1999): 83.
15. Stenson and Gray (1999): 93–114.
16. As Rosenthal notes, it is widely recognized that the most complex issue in this area is the identification of beneficiaries. Benefit sharing rests upon the pre-condition that we are able to demarcate the community that will be the beneficiary of such a bargain (see Rosenthal, 1997: 260).
17. Look, for example, at the case of copyrights where moral justifications, especially those based on the Lockean theory of creating value, dominate the grant of the right.
18. See Hirschleifer and Riley (1995); also see May (2000: 51) who notes that "If the provision of the building blocks of knowledge involves no necessary diminution of utility to the previous users/owners when re-used to create or invent, then the rationale for charging for inputs is not particularly robust. The marginal cost is close to nil – once an idea has been had there are no extra costs in others re-thinking it. Neither is it clear why protecting a particular creator (the idea's current possessor), over and above the creators who contributed earlier through their ideas is legitimate."
19. Law and economics literature explores, in many cases, alternate incentive effects that could lead to a grant of intellectual property right for the production of information. See, for example, Landes and Posner (1989); also relevant in this regard is Calandrillo (1998) who presents a detailed analysis of alternate incentive effects for the production of information.
20. Equity should no doubt be a strong basis for such a right, but trying to derive a right mainly based on fairness and equity considerations, wherein the maximum good for the maximum number of people is the aim, would theoretically work but practically yield little due to difficulties in implementation.
21. Marguiles (1993) has argued for the grant of a right to traditional knowledge in conjunction with the incentive to preserve biodiversity.
22. "Technological information" simply denotes the category of information that is mainly

produced through R&D activities (Zeckhauser, 1996: 12743). Also see Mandeville (1998: 46–47) in this context.

23. Another very interesting example of changed industrial organization due to cumulative innovation is the field of computer technology. For a detailed analysis of cumulative innovation and its repercussions in this field, see Scotchmer (1999).
24. In the same context, Scotchmer (1999: 1) notes that the "cumulativeness of innovation brings to the forefront four kinds of characteristics which were hitherto not taken into account by economics of information, namely, (a) later products can be improvements of earlier products; (b) later products can bring in cost reductions for the production of earlier products; (c) later products can be applications of earlier basic technologies; and, lastly, (d) later products can be enabling technologies such as research tools".
25. Saviotti explains the knowledge intermediation role played by dedicated biotechnological firms and public research centres for large diversified firms in detail (see Saviotti, 1998: 33).
26. Rosenberg (1982: 5) notes: "Where Arrow and Nelson's neoclassical theory of R and D has been criticized in the modern context as having portrayed research only as an activity resulting from an investment decision made by the profit-maximizing firm, the critical element for the investment decision being the returns on the investment. Nelson and Arrow argue that since social returns to research investment exceed the private returns faced by the individual firm, this would lead to under-investment by the firm (from a societal point of view) in R and D, is seen as having resulted from the micro-economic bias of treating the firm as 'black-box' as such." See also Foray (1995: 84–85) who notes vehemently that, "knowledge possesses particular properties (non-rivalry and cumulativeness) which leads us to observe that the value of a given item of knowledge is basically determined by the extent of its use and distribution within the system ... This effect is not confined merely to the application of existing knowledge in the production of conventional commodities; it extends also to the use of information to produce more information."
27. Foray proposes three models of innovative activity in order to enable us to distinguish the changing nature of innovation and its specific policy demands (see Foray, 1995: 89).
28. Elisabetsky (1991: 9–13) cited in Cox (1995): 10.
29. David (1993: 27) cited in Foray (1995): 80.
30. Cox (1983) notes that a survey of Samoan ethnopharmacopoeia revealed that 86 per cent of the plant species used in Samoan traditional herbal medicine shows pharmacological activity in broad *in vitro* and *in vivo* screens. Also see, Anderson (1986a), where he lists the 121 different species of plants used by the indigenous tribe of Akha, and Anderson (1986b), where again 68 species of plants in use by the Lahu people are listed, among many others.
31. Balick (1990: 26–28) notes that when compared to a 6 per cent activity level found in plants collected randomly, the activity level in plants collected through ethnobotanical surveys rose to 25 per cent.
32. For example, Vlietinck and van den Berghe (1998) point out potent anti-viral activity being confirmed in R&D centres from traditional medicines, research based on which is right now at the stage of clinical trials. Also see Raman and Skett (1998) who talk of validity of scientific studies which have shown that activities of plants to be antidiabetic when their traditional usage was also antidiabetic.
33. See Rausser and Small (2000: 177–179), where they discuss industry behaviour that seeks out biodiversity "hotspots" and ethnobotanical information.
34. Common property resource management literature has discussed the impacts of heterogeneity that can arise in such otherwise homogeneous groups as a result of exchanges with the external economic environment (see for example, Kanbur, 1992).

35. According to Ellickson (1994), "members of a close-knit group develop and maintain norms whose content serves to maximize the aggregate welfare that members obtain in their workaday affairs with one another". Ellickson (1994: 167) also cites other authors who have advanced similar propositions that are of interest for us, see, for example, Edwin (1977) *The Biological Origin of Human Values*, p. 362, who analyses how norms of cooperative behaviour evolve in small tribes.
36. For example, in Indian villages where a lot of property was managed in common, persons of the village community who disobeyed rules were punished with expulsion (Bromley and Cyrene, 1989).
37. See for instance, Schaaf (2000: 325–244), who explains how the system of preserving areas with special ecosystems by denoting them as "sacred sites" only worked through access restrictions.
38. Also see Chapela (1997) for a discussion of Mexican communities in Oaxaca that use internal management techniques to exclude outsiders and to implement sustainable biodiversity prospecting partnerships designed by them to further conservation and sustainable use of biodiversity held on their lands.
39. An experimental project based in Ecuador and being supported by the Inter-American Development Bank (IDB) seeks to transform indigenous knowledge into trade secrets. These are to be contracted away by way of material transfer agreements. For details, see Dutfield (2000): 89; Vogel (1996): 16–17.

6

The scope of the right to regulate access

Economic theory predicts that in the absence of property rights and regulative intervention, biological resources exhibit properties of a common good. Undesired use of biological resources by third parties becomes very costly to exclude, thereby resulting in over-use, what is also called the "tragedy of the commons". But the regulation of access as provided for under Article 15 of the CBD and its application to bioprospecting raises several important questions. This chapter picks up from the legal analysis of the right to access in Chapter 3 to deal with these questions.

Setting up access institutions at national levels that can perform all the functions that have been identified in Chapter 3 is a costly exercise. Most countries worldwide, especially developing countries, have not conducted inventories of their biological diversity. Although some countries like Australia, Namibia, and Costa Rica are in the process of developing detailed national biodiversity inventories, most other countries have a patchwork of databases and publications on aspects of biodiversity as their national inventories (ten Kate and Wells, 2001: 20). They also have very limited financial, human, and institutional resources to devote to such an exercise. In the light of this, the questions that assume extreme importance are: when can access regulation for bioprospecting be justified per se? And under what circumstances can bioprospecting provide the incentive for countries to incur costs in setting up elaborate institutions to regulate access to genetic resources?

Source countries' investments in setting up access laws and institutions for bioprospecting and/or charging firms for access to genetic resources can be justified only when two sets of conditions are satisfied. Firstly, it needs to be ascertained that conservation efforts as undertaken by national access authorities will aid present and future bioprospecting. Internalization of externalities through regulation is a costly activity, and therefore whether genetic resources have a positive economic value beyond conservation of species that call for such investments enabling scientific or commercial bioprospecting has to be evaluated.

Secondly, regulating access to genetic resources and charging firms to pay for biodiversity conservation can be defended only when it can be shown that bioprospecting itself is an activity with potential undesirable impacts on sustainable use and conservation of genetic resources. From an economic perspective, if bioprospecting does not cause externalities in biodiversity use that need to be internalized, then access is a pure tax on certain activities (in this case, bioprospecting) within a country where the users of genetic resources have to pay for biological conservation although they are not in any way directly responsible for its unsustainable use or extinction. In this case, the argument is not biodiversity conservation, it is just a plan to generate revenue for the country – similar to having an oil mine on one's own land. At best this will result in redistribution of profits from bioprospecting firms to indigenous and local communities and biodiversity support programmes when the access authorities function in an accountable and transparent manner. At worst, it will end up being rent-seeking behaviour of a state, through the exercise of its monopoly power on access rights, to skim off the economic surplus of firms that engage in drug R&D. Is this what the CBD imagines?

If the answer to the aforementioned conditions is in the affirmative, one needs to then investigate how the right to access can help solve the contractual issues that parties face in the imperfect market for bioprospecting and help capture the benefits of valuable genetic resources and ethnobotanical knowledge.

The focus of this chapter will be to answer these questions and to derive a property right for access to genetic resources that can be worked at the national level. Section I focuses on the potential value of genetic resources in general, and investigates as to whether market-based incentives for conservation do exist. It also assesses the extent to which biodiversity prospecting can cause externalities in biodiversity use. Section II evaluates different property rights options that can be considered in the case of access rights, and proposes the optimal definition for the right to access. The remaining analysis in this chapter focuses on identifying the conditions under which access can solve both responsibilities of biodiversity conservation and market creation for bioprospecting. In this endeavour, a focal point of the analysis is the feedback loop between market revenues that can be generated through activities like bioprospecting into biodiversity conservation.

I. REGULATING BIOPROSPECTING: THE CONDITIONS

From a review of existing literature it appears that legal scholarship has focused more on the form that the access right can take rather than on the basis on which the right can be justified, its contents, or the conditions that such a right needs to fulfil. As a result, although there have been several suggestions regarding rights structures, the question of how the access authority ought to perform its duties or its role in bioprospecting contracts has not received much attention.[1] A big exception in this regard is Krattiger and Lesser (1994) who focus on the role of the access right as a contractual facilitator in the market for bioprospecting rather than its outwardly attributes. They propose a proactive "brokerage" service under the name of the "facilitator" for access institutions. The "facilitator" is supposed to perform the role of bringing potential partnerships together in a manner consistent with the provisions of the CBD, including support for the exchange of necessary information to identify potential partners, advising on appropriate terms of contracts, and possibly underwriting some deals (Krattiger and Lesser, 1995: 212–213).

On the other hand, economic literature on the topic is largely devoted to aspects of valuation. There are many who focus on trying to pin down the economic value of biodiversity in general,[2] whereas others try to derive the loss that biodiversity extinction causes to medicinal research (see, for example, Principe, 1990; Aylward, 1993). Although the latter category could have been especially useful in clarifying some contentious issues, the problem is, as Artuso (1997a) notes, that the net present value to any pharmaceutical firm in obtaining rights for such plants varies largely depending on the assumptions of the calculations therein.[3] There are very few studies that focus exclusively on the conditions under which bioprospecting can actually provide the right incentives for conservation or on the conditions which need to be met for this market to develop.

Bioprospecting as a successful economic venue

Bioprospecting can serve as a market-based incentive for conservation only when it can be proven that genetic resources are non-substitutable inputs for one another in the drug R&D process (Rausser and Small, 2000). If this is true, then the expected benefits of conservation will exceed the opportunity costs of holding these assets because firms will be ready to pay for genetic resources that have potential to be leads in the drug R&D process today or in the future (ibid.). The extinction of such resources symbolizes lost opportunities that could have otherwise been utilized. The R&D option value of biodiversity provides another market-based argument for biodiversity conservation.

Genetic resources: marginal or non-marginal inputs to drug research?

It has been argued that the value of an additional genetic resource is marginal to the R&D process thus making it insufficient as an incentive for biodiversity conservation (Simpson, Sedjo, and Ried, 1995).

Simpson, Sedjo, and Ried developed a model of demand for biological diversity for use in drug research on the assumption of a declining marginal value of biological samples to the drug industry. In their model, the expected value of screening additional samples decline due to the increasing probability of having satisfied the demand for new drugs from samples already screened. The model relies on two critical assumptions. The first one is *uncertainty*; that is, it is unknown prior to testing as to whether a given lead will generate a potential discovery or not. The second assumption is one of *redundancy* among the leads; that is, since any given discovery can be made only once, potential leads that contain the components of similar discoveries are of course superfluous. Based on these two assumptions, they argue that the search for species providing a certain type of benefit will terminate with the discovery of the first drug from that particular species. As a result, in their model, the value of the marginal species declines rapidly if the actual success rate of natural products research differs from the level necessary to maximize the value of the marginal species. Because, in such a case, the willingness to pay amongst private users to prevent the collection of leads from becoming somewhat smaller will be marginal. From this they conclude that although biodiversity is valuable in itself, the value of an additional genetic resource is marginal to the R&D process thus making bioprospecting an insufficient incentive for biodiversity conservation (Simpson, Sedjo, and Ried, 1995: 166).

But these assumptions by Simpson, Sedjo, and Ried do not reflect the reality of the drug R&D processes. The model finds rebuttal in the behaviour of the drug industry. Rausser and Small (2000) prove through evidence based on the behaviour of the drug industry and a model of applied research which segregates leads based on their expected quality, that a market for genetic resources will evolve because genetic resources as found in tropical ecosystems are non-substitutable. They note that:

> if genetic resources were absolute substitutes of one another, then we must expect to find bioprospecting projects spread evenly around the world, even apart from observing behavior such as locating collection sites in close distances to R&D

> labs, ease of access among other things – but this is in fact not the case. Bioprospecting firms have indeed shown extreme willingness to embark upon expensive collection efforts in remote but ecologically distinctive locations including under seas, to get admission to which places they commonly spend many staff hours negotiating access contracts with host country governments and their agents (Rausser and Small, 2000: 179).

They use the behaviour of the various actors in the drug industry not only to address the more empirical issue of whether informational rents associated with genetic resources might be large enough to play a significant role as a source of biodiversity conservation, but also to elaborate upon the circumstances under which genetic resources may qualify to be called components with such informational advantages.

The critical thrust for their argument comes from the fact that genetic resources are not perfect substitutes for one another, especially those that show unusual promise for the drug discovery programmes. And it is these kinds of genetic resources, according to them, that could "command informational rents that reduce the costs of such research" (ibid.: 173). They find support for their argument in the behaviour of the drug industry which suggests that bioprospecting is a targeted and not a haphazard activity, wherein scientists and firms can draw on rich bases such as ecology and taxonomy to narrow down on genetic resources that show unusual promise.

This intriguing finding based on the behaviour of the drug industry has reinforced the relevance of access regulation and access institutions for bioprospecting. The largest number of undiscovered yet "promising" species exist among higher plants, invertebrates, and microorganisms that possess unique physiological characteristics and thrive in extreme environmental conditions (ibid). Bioprospecting can generate sufficient revenues for setting up access institutions in the more exotic geographical zones which host exotic species, or the so-called "hot spots" of every country, where the non-substitutability condition is satisfied.

The R&D option value of biological diversity

The quasi-option value of biodiversity is another convincing argument that has been pursued at the microeconomic level to argue for market-based incentives for conservation. A quasi-option value means the value of retaining the option to make a irreversible investment by postponing the decision to a later point of time. For example, when a government in a source country makes the decision to convert forest into agricultural land, it makes an irreversible decision (see Artuso, 1997a). It gives up the possibility of waiting for new information to arrive that might affect the desirability or timing of the expenditure; it cannot disinvest should market conditions of agricultural products change unfavourably in comparison with forestry products (Pindyck, 1991: 1112). The quasi-option value is the value of keeping this option open, by postponing conversion decisions to a later point of time so that benefits can be derived from future options that may be far more valuable than those currently known (Swanson, 1995). In the case of bioprospecting, where the actual value of biological resources and specific ecosystems is unknown, the value of maintaining the opportunity to benefit from a protected site will rise if bioprospecting opportunities become more profitable in the future. Subsequently, as Artuso notes, if the decision to develop or preserve a natural habitat can be postponed until new information (e.g. new screening technologies, results of present bioprospecting activities) is received, then a correctly specified decision analysis would include a calculation of the expected maximum benefits of the economic opportunities that will become available in the future, given current land use choices (Artuso, 1997a: 74).

Although much of the option value literature focuses on finance, looking at issues of biological diversity in terms of quasi-option values is not new.[4] Usually, quasi-option values are always positive (or zero if the exercise price of the option right exceeds the price of the underlying real asset – biological resources here – minus the exercise price of the option), but there could be a problem in determining beforehand whether it would yield higher profits to wait for newer information (Swanson, 1995).

Despite this, it has been suggested that quasi-option value of biological resources could generate enough incentives for access institutions within countries. A very good proposal in this respect has been made by Artuso (1997a). He argues that the quasi-option value associated with bioprospecting can be made use of by source countries if they focus on how new information could affect the value and number of species that are not adequately protected within the country. With numerical examples, he explains how (a) new medicinal research and the disturbing occurrence of other drug-resistant diseases, and (b) the development of the source country's own biotechnological capabilities to exploit its biota, act as two main factors that increase the value of the species in such biodiversity-rich sites within the country (ibid.: 75).

Exploring the link between bioprospecting and sustainable use of genetic resources

The most direct link between intellectual property rights as granted by Article 27(3)(b) and the issue of sustainable use and conservation relates to the degree to which a species should be exploited before one could term it "unsustainable" (Dhillon and Amundsen, 2000). The collection methods for species used by bioprospectors, coupled with a specific nonchalance about the vulnerability or other characteristics of species to be harvested has led to the view that bioprospecting may not, after all, constitute a sustainable activity.

As already discussed in Chapters 2 and 4, the two stages of the pharmaceutical drug R&D process that lend support to indiscriminate harvesting/collecting of plant species are the pre-clinical and clinical testing stages and the large-scale production of drugs (Cragg et al., 1995). Since huge amounts of raw material are required to produce minimal quantities of the drugs in certain cases, there is a very high probability that species whose medicinal value is discovered are harvested/collected indiscriminately from the wild.

Although it is still not clear what percentage of total cases of drugs discovered from natural substances have led to the unsustainable use of the genetic resources that form the core substance of the drug, several examples exist. Two cases that are repeatedly mentioned are those of the Rosy Periwinkle and the Himalayan Yew. The plant Rosy Periwinkle that is required for the production of the drugs containing vincristine and vinblastine is said to require two tons of dried leaves to produce just one gram of active medicine (Dhillon and Amundsen, 2000: 110). It is also reported that in the early phases of production, when the harvesting of Rosy Periwinkle created extreme pressure on naturally occurring amounts of the plant in Madagascar, the firm shifted to India and Mozambique for supplies. After a while, given the large quantities of the raw material required for production of the drug, the cultivation of the plant was the only solution left. Nowadays, Madagascar cultivates the plant for Eli Lilly and supplies around 1,000 tons per annum for the production of medicines to cure childhood leukaemia (Dhillon and Amundsen, 2000: 110), but in regions where the plant used to grow naturally abundantly, the species is hardly to be found due to unsustainable collecting prior to cultivation (Wood Sheldon et al., 1997).

The case of Taxol® also shows the extreme amounts of raw material required to test a potential source of pharmacological activity at the clinical trial stages, even before the developers can be sure as to whether a drug will accrue from the process. As ten Kate and Laird (1999) and Cragg and Boyd (1996) note, preclinical and early clinical studies on Taxol required approximately 900–6,800 kilograms of Pacific Yew (a close relative of the Himalayan Yew) bark to yield a total of only about 1.3 kilograms of the drug. As clinical trials advanced, however, the annual requirements leapt to approximately 27,250 kilograms of bark per year, creating a supply crisis.

Some other reported cases of such over-collection as a result of discovered medicinal value include *Prunus africana Rosaceae* (active compounds: sterols, triterpenes, and n-docosanol), *Pilocarpus spp. Rutaceae* (active compound: Pilocarpine), *Podophyllum hexandrum Berberidaceae* (also called Himalayan Mayapple, active compound: podophyllin) and *Ravoulfina Serpentina Apocynaceae* (also called Indian Snakeroot, active compound: reserpine).[5]

In most cases of this kind, the market price at which the raw genetic material is traded does not reflect the opportunity costs of the resource, thus resulting in externalities. For example, in the case of Jaborandi, which was discussed in Chapter 4, when the medicinal use of Pilocarpine (the active compound from the shrub Jaborandi) was discovered, the shrub was almost harvested to extinction in Brazil and Paraguay.[6] At that time, the wholesale price paid for a kilogram of the dried plant material was US$0.28, although in 1989, the estimated sales of Pilocarpine in the US market alone reached US$28 million and more than 1,300 tons of the plant material were being imported annually. E. Merck maintained throughout that their in-country suppliers supported only sustainable forms of harvesting from local collectors, but this could never be objectively verified as, by the end of the 1980s, the wild collections of the plant became almost extinct. It is not clear as to how much and how far the policies of E. Merck, and the company's problems with the source countries in ensuring reliable supplies, contributed to this. In many such cases where a commercial use is discovered, there is a real threat that the hype associated with the plant's economic value may lead to its indiscriminate collecting and storing, which may or may not be caused by the acquisition policies of the firm developing the medicinal product, as noted in Chapter 2. It is because of these cases that authors such as Aylward (1996: 222) note, "The protection and maintenance of biodiversity should be included in the economic analysis of biodiversity because the supply of this input requires a very significant social cost".

Despite the presence of such cases, it is not possible to generalize the impact of drug R&D on genetic resource use and conservation. There are several reasons why a discovered medicinal use leads to such indiscriminate use. Firstly, whether raw materials are required for the manufacture of the drug after it is developed or not depends on the kind of drug in question. To recap the discussion on this issue from Chapter 2, nature may either provide a source of the drug itself, or may provide a simpler precursor which is readily converted to the drug through a semi-synthetic process (Cragg, 2004, personal communication). Of the several categories of drugs that use natural products as the starting point – biological, natural product-based drugs, semi-synthetic drugs derived from natural products, or totally synthetic drugs – biologicals and natural product-based drugs are usually produced naturally and require inputs of the genetic resource in question (ibid.). But often given the complex structures of natural products, while they may be synthesized in the laboratory, the development of an economically viable synthesis is not possible, and therefore there are also instances where drugs that are derived from natural products depend on raw material for their production (ibid.). A good example of this is the anti-cancer drug topotecan. Although topotecan is a synthetic derivative of the natural product, camptothecin that is found in a Chinese plant, the production of the drug still begins from camptothecin (ibid.). Chinese and Indian pharmaceutical organizations originally supplied Smithkline Beecham with natural camptothecin for the manufacture of topotecan between the mid-1980s and the early 1990s. Currently, a different international broker is supplying the raw material that it sources from Brazilian plantations (ten Kate and Wells, 1998: 6).

Even where artificial synthesis of the compound may be possible, it does not make economic sense for the firms to invest in developing alternate sources of supply until it is evident that a commercially viable drug can be developed from the compound (as discussed in Chapter 4). Until then, the two earlier stages of the R&D process that require huge quantities of the raw material will have been completed using other ways of sourcing.

But even in cases where a profitable drug has been developed, if firms can get the raw genetic material from the wild through contracted collectors or the spot market traders, it may still be a preferred option from the firm's perspective, because it is a cheaper alternative than developing infrastructure to produce the genetic material through other means (Wood Sheldon et al., 1997). The individuals who supply directly or through private profit brokers are often the poorest in the countries where the plant with the discovered medicinal use grows naturally. As a result, the raw genetic material is sold on so-called "spot market" prices that do not reflect the opportunity costs of the resource or the externalities created by such harvesting and collecting.

Similar problems exist in the case of botanical medicines, too, and maybe to a greater degree, since botanical medicines are mostly total plant extracts (Laird, 2004, personal communication). In the botanical medicine industry, raw materials are sourced as "bulk commodity from a long line of intermediaries" (Pearce and Laird, 2003). As a result, companies are usually unaware of the country of origin of such material, their production mechanisms (whether cultivated or harvested), its impact on sustainable use and conservation, as well as on livelihoods of people who collect or harvest the raw material (Laird and Pearce, 2002; Pearce and Laird, 2003).

An industry survey of the botanical medicine industry conducted by Laird and Pearce (2002), which interviewed 54 companies (17 from the United States, 26 from Europe, 7 from Latin America, 2 from Asia and 2 from Africa), investigated attitudes towards raw material sourcing, among other issues. They noted on the issue of cultivated versus wild-harvested material that):

Well over half of the companies responding to this question answered that the majority of their material comes from cultivated sources (however, further discussion with some informants revealed that many had differing definitions of "wild" and "cultivated"). Twenty seven of these reported only that more than 75% of their material was cultivated (the average was between 60–90%). Only 9 companies reported larger portions of wild harvested material, and almost all of these were based in countries with large raw material supplies available in the wild, usually in developing countries (e.g. Ghana, Brazil, Peru and some U.S.). Eight companies replied that they had "no idea" of the breakdown, a response that would probably have been appropriate for a larger number than 8 of the respondents. In general – with the exception of raw botanical supply companies, and interviews with purchasing agents – this appeared to be a question companies had not considered, and the answer to which they would be hard-pressed to find. Most companies buy material via one or more intermediaries, and the manner of production or collection is not often evident from accompanying paperwork (Laird and Pearce, 2002: 19).

In the same survey, when asked if sustainability of the raw material for the production of a product was a factor that motivated decisions on product development and marketing, on an average, the larger companies tended to agree, whereas for the smaller ones, "sustainability and fair trade were clearly additional costs and issues they were not willing to take on board in any significant way" (ibid.: 23).

II. INSTRUMENTS AVAILABLE TO INTERNALIZE EXTERNALITIES

Although the medicinal use of such species is the cause for their over-harvesting and extinction, it is predominantly the legal vacuum in source countries that allows parties to trade with resources that they neither hold nor are accountable for, thereby generating externalities.

Therefore, in order to prevent a potential or actual medicinal use leading to unsustainable use of the genetic resources in question, the right to access has to eliminate incentives amongst users to treat such biological resources as a commons or open-access resource. The problem can be resolved with the help of a strictly enforceable access regime which would help the market price of genetic resources reflect their costs of extinction by the establishment of mechanisms of either property rights, taxation, or even detailed regulation, on the basis of the broad guidelines of the CBD. Legal literature has proposed several options for a property rights' regime on access.

One solution that has been proposed is the introduction of extremely stringent access laws. Such laws mainly aim at regulating the behaviour of the external (here, mostly foreign) firms vis-à-vis biodiversity depletion. Whereas, on the one hand, the laws do include stringent requirements with regard to the revelation of research details, considerably less attention is paid to harvesting/collection techniques that may be used by collectors and bioprospecting firms. A major shortcoming of these attempts has been that the mechanisms that the laws prescribe are not linked to causation. Since the background information about the kinds of impact is usually harder to get grips with, the laws set up a user-fee system, whereby a general amount of compensation is to be paid by each and every bioprospector, irrespective of the kind and amount of externalities their activities may create.

Some other proposals have recognized that just authorizing the national access authority to ask for details is not sufficient to control the problem. They propose that source countries take a proactive role in setting up contractual mechanisms for raw material supplies. Since raw material supplies are not replaceable inputs in the R&D process, reliable supplies is a concern for firms when entering into such contracts (Wood Sheldon et al., 1997: 7). The notion of "pharmaceutical crops" is one such suggestion to deal with the issue of over-harvesting (Cragg et al., 1995). Daly (1992) proposes cultivation programmes with the full participation of the national government as a potential solution. He cites the *Ancistrocladus korupensis* example to demonstrate how collection and cultivation programmes can be taken up by bioprospecting institutes and national governments to address the problem of over-harvesting of wild species.

Such solutions may suffice where collectors and harvesters of raw genetic material are the firms and institutes themselves. But, in practice, as a range of individuals and organizations can collect the material and sell it on the spot market, solutions have to be able to cater to two levels. In the case where the firm itself is the contracting party, the inclusion of stringent rules mandating details of research can be an effective check to ensure sustainable use. Collections by firms can be controlled to a greater extent with the help of a legal framework that mandates the revelation of information related to the intensity of the harvesting and collecting, thereby allowing the access institutions to keep a check on sustainability.[7]

Regulating spot market trading of genetic material, on the other hand, is much more difficult. Unsustainable use results because no one, including the source country government, is able to assess the number of parallel efforts that may be going on to harvest/collect any given medicinal plant species, thereby eliminating the possibility of ascertaining and preventing the potential harm of such an activity *ex ante*. Stringent laws per se do not provide an adequate basis for the national regulatory authority to detect and hold such collectors individually responsible for the extinction, since that would require massive monitoring capabilities. To solve this problem, legal frameworks have to be able to eradicate the profit margin in trading with such resources.

Theoretically, there are several forms that the right to access could take. One solution is to define property rights to promote market solutions (see generally, OECD, 1999). It is also possible to imagine a tax-like mechanism – where a tax is set at a price that equals the spot market price plus the loss due to externality, it will be less lucrative for third parties to indulge in trading in such resources. There is also the possibility of zoning and regulations.

Since different economic instruments of internalization will have significant and differing administrative costs related to establishment and enforcement, a comparison of the costs and benefits of each of these instruments is needed to choose one option over all others. The economics of natural resource management has dealt with many aspects of comparison, which are taken up below.

Property rights

Private ownership, to the extent that it allows for a single owner to manage such resources, will ensure control over how many people access the resource and also the price at which they access it. Since the private owner will bear the full costs of exploitation (and eventual extinction) of the resource, the price at which other users will be allowed to access the resource will reflect the total costs of such access.

This insight is owed to Coase's contribution to the problem of social cost. By using examples from the realm of renewable resources (like grazing, water, noise and air pollution), Coase argued that when there are no transaction costs, absence of wealth effects, and when property rights are well-defined and enforced, Pareto-efficient allocations will result when parties internalize externalities through bargaining (under competitive conditions). In this setting, liability rules prescribe the bargaining thresholds for parties to negotiate internalization (see Coase, 1960).

In practice, it is the courts that enforce liability rules when rights are infringed. This may not be possible in the case of genetic resources for many reasons. Firstly, in order to be able to enforce compensation payments based on liability rules, courts have to be able to find and settle upon the market price for the infringed genetic resources. Finding the "right" market price is as difficult in the case of genetic resources as with other precious renewable resources worldwide. As the previous section on valuation of genetic resources has also shown, although economists and ecologists alike have been able to affirm the positive value of genetic resources, it is very difficult to settle upon the precise foregone losses of depletion or extinction of such resources.

To settle upon foregone losses, valuation criteria are required that take into account the "non-use" value of genetic resources or an estimation of their net present value for pharmaceutical research.[8] Both these options are difficult, since there are no market estimations of non-use values of genetic resources, either direct or indirect (Brown, 2000), and there is no consensus on the net present value of medicinal plants for the drug industry (see, for example, Principe, 1995). As a result, court authorities will be faced by a lack of information on estimating damages when private property rights based on such genetic resources are infringed upon by third parties.

Secondly, well-functioning property rights supported by liability rules have the effect of deterring infringements by third parties when there is a possible likelihood of legal action and of claiming damages for the harm done. This compels parties indulging in certain activities to assess the costs of their activities and the risks associated with them. But in the absence of well-functioning legal systems that enforce compensation payments without delays – which is the case in most developing countries with heavy litigation backlogs – the deterrence effect may only be marginal. As a result, trespassers may have differing threat points in assessing the costs of their activities and deciding their course of action.

Lastly, for property rights to result in efficient internalization, as in the Coasean situation, bargaining must occur under competitive conditions. If property rights are well-defined then by applying the Coase Theorem it must be that regardless of whether the firms or the communities or other private parties get the rights to the genetic resources, the ultimate allocative result remains the same. This is because the parties can always bargain between each other and reach the stage of efficient internalization of externalities. However, this cannot be achieved if the bargaining has to take place in imperfect markets with high transaction costs, which is the case in bioprospecting.

Taxation

A tax can reduce harmful effects by setting a price equal to the external costs of exploitation that has to be paid by each user. But setting the price equal to the external costs of exploitation poses the same difficulty as that of finding the market price of infringement of property rights for genetic resources. Since it is very difficult to settle upon the precise foregone losses of depletion/extinction of genetic resources, setting the tax equal to external costs of exploitation is not possible. It can only be a best guess or an approximation. Therefore, designing the access right as a user-based fee will not result in efficient internalization, but rather will be a best guess on how harmful bioprospecting might be reduced.

Zoning and individual transferable quotas

Renewable resources are subject to periodic acceleration of growth or depletion and therefore have to be studied in an inter-temporal setting (see Brown, 2000). The characteristic of growth differs between different kinds of renewable resources. For example, population is the organizing unit for fish, whereas for trees, the optimal duration of rotation (that is, how long it takes to grow back once it has been felled) is the organizing unit (Brown, 2000: 888–889). Furthermore, resources within natural ecosystems provide services that are linked through complex interdependencies. That is, the species in one ecosystem are dependent on species and reproduction cycles in other ecosystems through networks of mutual interaction. As a result, depletion of one set of resources may actually impact upon many more resource settings than one might imagine. This characteristic is also referred to as the "spatial dimension" of renewable resources in environmental resource economics (e.g. Brown, 2000). Property rights allocations on renewable resources have necessarily to take spatial impacts on board in order to be effective.

Traditional economic solutions that propose the restriction of access either through private property rights or taxation systems neglect the inter-temporal and spatial dimensions of renewable resources. Restriction of access in these ways will merely slow down the process of extinction as the harvesters exploit the unconstrained inputs.

Economics of environmental resource management suggest a different policy tool that takes into account the spatial and inter-temporal dimensions of renewable resources more effectively. This solution is the creation of property rights by the state and their distribution to harvesters. These property rights are rights for the harvest of a specific quantity and quality of the resource, but only for a specific duration.

These rights are called individual transferable quotas (ITQs). ITQs allow for uncertainty in the natural provision of these resources to be taken into account – that is, sometimes nature may recuperate faster and sometimes slower. Presently, ITQs are being used very successfully in the realm of fishing.

In the case of genetic resources, the specific tropical ecosystems that fulfil the non-substitutability criteria and have to be regulated from unsustainable use are mostly already demarcated (or zoned) naturally. Regulators only have to identify these zones and create a market for ITQs for genetic resources within the zones. As always, the crux of making ITQs functional are the rules that monitor them. The monitoring of inter-temporal dimension depends on setting the right characteristic to measure growth of the resource. Similarly to the fishing industry, where the organizational unit is population, appropriate organization units for tropical ecosystems have to be set out by national policy makers. This requires expertise from specialists such as botanists and environmentalists in the regulatory processes. Protecting the zoned ecosystems, from a spatial dimension, is also demanding. The rejuvenation of resources in the zoned ecosystems may be affected adversely due to exploitation of resources in ecosystems that surround them. For this purpose, national regulators have to demarcate "buffer zones" around these protected zones so that spatial effects can be dealt with effectively.

If the rules for ITQs are set in this way, national access authorities can regularly monitor the status of resources within sought-after ecosystems in the country, and trade rights to harvest/collect specific quantity (and quality) of genetic resources for a specific duration only. When the pressure on the ecosystems increases, or when species in demand show signs of reaching unsustainable use status, the quotas should be cancelled and the ecosystems should be allowed to recover.

An ITQ system coupled with high fines for collection activities conducted without legal ITQs can also effectively control the tendency of third-party trafficking in genetic resources harvested or collected without quota permits. A word of caution on ITQs though is that they rely on very meticulous enforcement systems for their success. This may be a problem in the case of developing and least developed countries. A framework for implementation of the ITQ system for access has been proposed in Annex 1.

Creating a market for access and traditional medicinal knowledge

Several results stand out from the analysis set out in the preceding section in this chapter. Firstly, access regulation for bioprospecting can be justified on grounds of biodiversity conservation since, in many instances, discovery of medicinal uses of genetic resources leads to unsustainable exploitation. But to tackle this, setting up a national biodiversity strategy and identifying local priorities for sustainable use and conservation are actions that need to be taken before any national access law is devised (see Chapter 3). Secondly, conducting country studies, devising action plans for the national biodiversity strategies, and setting up access laws and institutions to implement the mandate in the context of bioprospecting requires elaborate investments on the part of source countries. But the market has its own limitations, bioprospecting can act as a market-based incentive for biodiversity conservation only in respect of those genetic resources that fulfil the non-substitutability criteria, and this may not be the case with *all* flora and fauna within a given source country. But in ecosystems where the non-substitutability criteria is in fact fulfilled, national governments can generate revenues for their conservation through zoning such areas and charging for bioprospecting taking place today, or even harnessing option values within these areas. This is not to say that conservation programmes in other parts of a country should not be pursued – it is only that bioprospecting cannot be a market incentive for such conservation programmes. Finally, from the discussion in Chapter 5 on traditional knowledge, it follows that those communities which possess the relevant ethnobotanical knowledge that can contribute to drug R&D are, for the most part, also limited to those specific extraordinary locations and ecosystems. Therefore, zoning and regulation within these specific ecosystems would, to a large extent, overlap with such communities and cover such traditional medicinal knowledge too.

Legal rules for access

Legal rules for access have to deal with three critical issues: how PIC should operate; how access can increase legal certainty and thereby foster trust amongst contractual parties; and how it can represent communities effectively.

How should the prior informed consent (PIC) process operate in practice?

An accountable PIC process will be one that decides between multiple potential uses of genetic resources and grants "informed consent" only to those cases that put genetic resources to their best use. Deciding between multiple demands on a finite set of national resources, while keeping in mind their present and future potential (in terms of option values), requires an evaluation of investment opportunities in genetic resources. For example, the decision whether to preserve or irreversibly convert to alternative land use or allow bioprospecting in a given ecosystem should incorporate the information about expected benefits that will accrue in each one of these alternate situations (Artuso, 1997a).

This, in turn, assumes the presence of a well-functioning market for biodiversity uses within the source country. Based on the demand in this market, the national access authority can put the resources to their best use. In the absence of a well-functioning market, the national access authority has to mimic such an outcome in deciding whether or not a particular offer would put its genetic resources to their best use.

In practice, this system will look somewhat as follows. When the access authority gets an offer for bioprospecting, its main task will be to assess the competing uses, their comparative benefits and costs, based on the information of the country study and the national biodiversity strategy, in order to decide which resource use would yield the best outcome (Artuso, 1997a: 66). Specifically for bioprospecting, the critical issue is whether the request for access would actually lead to other potential uses of that particular geographical area to be altered, postponed, or even cancelled – and when the foregone benefits of these activities are taken into account, whether bioprospecting would still be lucrative (ibid).

Such a transparent and accountable procedure for granting PIC will help the market price of genetic resources to reflect the full costs of their use and extinction. When enforced, along with appropriate sanctions, such a regime can eliminate the incentives amongst its own citizens to "traffic" in genetic resources without legal consent. Such accountability and transparency in the PIC process will also enhance incentives amongst foreign firms/bioprospectors to access genetic resources only when legally permitted. In cases where they choose to do otherwise, tracing such unauthorized access will become easier, increasing the threat of action against trespassers.

Legal certainty and transparency in access

In addition to a clear procedure on which PIC should be granted, access laws should enhance legal certainty and transparency in order that trust and credibility in contractual relations is enhanced. There are several aspects to this.

Creating a good interface between four different kinds of property rights regimes through access regimes – access rights, patent rights, rights over private lands where the genetic resources may be situated, and community intellectual property rights – will depend on how clearly the roles and responsibilities of both users and providers are laid out. Failing this, false expectations of contracting parties may stall a process that has already commenced, or even hinder bargaining from taking place, as in the ICBG Maya case. As discussed in Chapter 4, the ICBG Maya case took a sudden turn in 1999 when several indigenous peoples' organizations called for the programme to be suspended. Some of the main problems had to do with the way in which the Mexican law on access dealt with proprietary claims, such as: Who could claim intellectual property on the research and its results? Could a group of Mayan healers claim rights on all Mayan traditional knowledge? How should benefits be shared with landowners from whose land genetic resources were used? (Berlin, 2004, personal communication).

Source countries also bear a critical responsibility in creating and transparently implementing the link between biodiversity programmes identified by the national biodiversity strategy and the funds collected by the national access authority through bioprospecting. Given the potential for political instability and lack of legal and administrative transparency in developing countries, access laws should consider incorporation of check-and-balance mechanisms for public scrutiny. Other measures that can go a long way in ensuring appropriate implementation of access laws are inclusion of public organizations in annual reviews, and scope for public intervention in cases where there is a lack of accountability by national access authorities.

Access authorities have to eradicate any scope for arbitrary decision-making in access granting in order to encourage firms to invest. Discretionary powers within access laws have to be curbed and basic legal guarantees of transparent procedure have to be spelt out. There are three ways in which the legal framework can provide direct help in this respect. First, the grounds on which access can be denied or revoked should be mentioned and enforced in an entirely transparent way. Secondly, access authorities should make available information that is used as a basis for decision-making and price-setting to the public.[9] This way, not only is the bioprospecting firm aware of the kind of genetic resources within the country, but it can assess the approximate factors that will form part of price setting. Lastly, a right to appeal has to be provided for in case access is denied for reasons that the firm/user did not intend or cannot change.

Minimizing bureaucratic costs of access

Many commentaries on industry problems in the market for bioprospecting point to the lack of laws or extremely bureaucratic laws as major impediments to successful bioprospecting contracts. Extremely bureaucratic access laws may mean that costs involved in following the access procedures make working of firms in these countries uncompetitive (ten Kate and Laird, 1999; Laird and ten Kate, 2002). In this context, ten Kate and Laird (1999: 298) note that "Companies are particularly troubled by what they perceive as a growing divergence between access measures introduced by policy-makers and what is feasible on the ground".

What is worse is that such elaborate laws may not necessarily enhance legal certainty. Reviews of newly enacted or existing access laws corroborate this. Ten Kate and Wells (2001) note that in China, there are at least seven governmental departments with jurisdiction to grant access rights there, whereas Australia has more than 40 different regimes for the granting of collection permits within the different states (ten Kate and Wells, 2001: 21). In the Andean group of countries that adopted Decision 391 on Access in 1996, three different situations exist in practice: whereas Bolivia has adopted implementing regulations, Peru and Ecuador are considering draft regulations, and Venezuela and Columbia have deemed Decision 391 to be directly applicable at the national level (Correa, 2003). Within these countries, the complexity of access procedures has created severe confusion and uncertainty regarding the rules and their implementation (ibid.: 805).

In some cases, confusion on the status of access laws also unwittingly promotes unsustainable use of genetic resources. Cat's Claw (*Uncaria tormentosa*) is a good example. As a result of confusion in the interpretation and implementation of the Andean Pact Decision 391 as to whether the collection of biological materials for commercial purposes other than as a source of genetic resources requires permits, such as their use in the botanical medicine industry, Peru has been permitting the collection of several plants that are used in the botanical industry, one of these being *Uncaria tormentosa* (Correa, 2003: 798). The government is also simultaneously encouraging the exploding demand for exports of Cat's Claw from Peru for use in the botanical medicine industry, since it provides an economic alternative for cocoa growers (Efe News Services, 1999, cited in Alexiades, 2002: 94). Such an interpretation of Decision 391 has led to great concern for the ecological sustainability of such resources in Peru (Alexiades, 2002).

Create a market for traditional medicinal knowledge

Defining a right to knowledge in the sense of ethnobotanical knowledge, as set out in Chapter 5, does not pre-empt contracting automatically. Access laws play an important role in augmenting the right to traditional medicinal knowledge. A major task of the representatives of the communities in bioprospecting contracts, is the resolution of principal-agent problems that may arise between the communities and the access authorities that represent them. Access laws should also augment laws on traditional medicinal knowledge by providing mechanisms that help communities overcome the potential shortcomings of community structures, such as caste, creed, or gender inequalities, which can limit their participation in bioprospecting contracts and conservation programmes.

Market traditional medicinal knowledge

The important role that ethnobotanical knowledge has to play in the drug R&D process has major implications for how such information can be harnessed. Assuming that there is knowledge that is beneficial, the cost involved for firms to obtain information about its existence and potential benefits may completely prevent such collaborations from occurring in the first place. Until now, firm data shows that in cases where firms have chosen ethnobotanical information as an R&D input, they have done so on the basis of information found in journals such as *Ethnobotany* and *Ethnopharmacology*.

The market for bioprospecting and traditional knowledge does not have an information-efficient equilibrium – that is, all relevant information is not reflected in the market prices. Contrary to an information-efficient market, where no agent has the incentive to acquire information on which prices are based, in this market there are particular informational rents associated with investing in such information (Grossmann and Stiglitz, 1980). This is especially true here because pinning down the interface between ethnobotanical knowledge and modern medicinal research is the crux of valuing such knowledge. Therefore, national access authorities should initiate programmes to validate the importance of ethnobotanical knowledge for modern medicinal research in order to capitalize on the economic value.[10]

This would involve investing in an upfront research effort in registering ethnobotanical knowledge and in correlating it to modern medicinal equivalents. The commercial value of such registers would be limited to those groups that possess a full body of species and their own set of ethnobotanical knowledge (as identified in Chapter 5). Even within such groups, creating effective registers of ethnobotanical knowledge presents real challenges for source countries. How does one document and record traditional medicinal knowledge adequately, especially since most of the uses are closely intertwined, multiple species are often involved in treatments, and the whole is embedded in complex social and cultural conditions? How does one develop effective benefit sharing across and within communities, and monitor these practices over time? These and other issues relating to the nature of market demand for traditional knowledge, cultural and social complexity, and equity need to be resolved satisfactorily if the commercial potential of such registers is to be exploited (Laird, 2004, personal communication).

It remains to be seen whether source countries themselves can discharge this function successfully, owing to the scientific nature of the activity as well as the required financial upfront investments. It is advisable that source countries create an enabling environment for such investments in registration activities to be carried out by other national institutions, public or private. Benefit-sharing, in return for such a useful service, would be a valid demand because it would be directly based on domestic value added. To prevent competing claims and undue expectations, access institutions would have to make room for distinguishing between the role of ethnobotanical knowledge – as the starting point of R&D – or the main information itself, on the basis of which the invention was based on, and set different packages for benefit-sharing accordingly.

The access authority as the agent of the communities also has to facilitate contracts for ethnobotanical knowledge. Ideally, this requires a monitoring mechanism that can check the value addition of the traditional knowledge-based information to the drug R&D process so as to ensure appropriate compensation, as discussed below.

Resolving the principal–agent relationship between communities and the access authority

The problems between the communities and their representatives/agents, the national access authorities, have a large potential to hinder *ex post* bargaining from taking place efficiently. This can be related either to the exercise of their rights on ancestral lands which they occupy and where firms are conducting collections of tangible genetic resources or to the use of their intangible traditional knowledge components.

This relationship resembles that of shareholders and managers in corporate transactions to a certain extent. If the access authority has to bring about the best results as the representative of the ethnobotanical knowledge of indigenous and local communities within its country, then there needs to be incentives in place that prompt the access authority to act as if it were a joint owner of this pool acting to maximize the joint value.

The main issues that affect the communities in this relationship are: (a) the agent's private information that makes it more difficult for the principal and the agent to agree upon terms; (b) the agent's lack of monitorability during the life of the contract. The solution to these issues lies in mechanisms that promote greater control of the principal (the communities) over the actions of the agent (the access authority) so that private information problems can be tackled, and greater control will give the communities stronger monitoring incentives.

There are many mechanisms that can help enhance greater control of communities over actions of national access authorities. One option is the provision of a *veto right* to communities, where communities can deny access to their lands and genetic resources if they deem this to be against their cultural or spiritual values. But such provisions pose the risk that a bioprospecting firm gets stuck in a contractual negotiation due to internal tensions between the communities and their government. This issue of internal tensions between the communities and their governments and the impact of lack of trust should not be under-estimated. Historically, since many such local and indigenous communities have been extremely marginalized by their own governments even small distortions can assume huge proportions.

Harnessing the potential of traditional knowledge for biodiversity conservation

Access institutions can also help support the incentives created by the right to traditional knowledge for *in situ* biodiversity conservation. As pointed out in Chapter 5, communities should be given the right to restrict access to their lands and genetic resources, in addition to their rights on traditional medicinal knowledge, so that common property resource management mechanisms can continue to be enforced. Access institutions should also include community participation in biodiversity conservation and use programmes, wherever possible. Through such programmes and other participatory mechanisms, access institutions should aim at promoting conservation efforts that integrate community techniques with the priorities identified by the national biodiversity strategy.

Access and the facilitation of contractual exchange

A facilitatory environment for contractual exchange means that the legal and contractual needs of all parties are taken into account, including the users. Although well-defined rights to traditional medicinal knowledge and access can solve many issues in the market for bioprospecting, the contractual problems that parties face while trading with information (traditional medicinal knowledge) as a good, as well as general issues of informational asymmetries, will still remain.

Information asymmetries and the role of access

The solution for assessing quality of goods to be traded in bioprospecting could very well lie in allowing screening by firms for optimal quality knowledge pools and signalling by communities of the quality of their knowledge pools (see Chapter 4).

Signalling through quality of legal frameworks

The supply of raw genetic materials or even ethnobotanical knowledge components for bioprospecting is not scarce. There are more than 80 tropical countries that are home to diverse ecosystems and local and indigenous communities worldwide. It is because these markets are competing for the same supply and the costs of signalling differ sufficiently amongst the "senders" that signalling could indeed work very well within this market. Source countries that have good quality pools and are serious in harnessing them through bioprospecting efforts can act as "senders" of the signal of the quality of the genetic material and traditional knowledge within their countries.

There can be good quality pools of genetic resources in a country, or bad quality pools. The biodiversity infrastructure investment that the CBD requires every country to make in terms of inventories of genetic resources and of ethnobotanical knowledge makes it possible for each country to have an estimate of the economic value of its own genetic wealth.

Let us consider a country such as India. Under the Convention, if it fulfils its mandate, it is possible for India to estimate the potential of the genetic resources within its territory. But this information is not known to an investor firm based in Europe or the USA. From an *ex ante* perspective, the probability for any firm of finding good genetic resources in India or elsewhere is the same, and is equal to the probability of finding bad genetic resources in these countries. Can India signal the quality of its genetic resources to potential bioprospectors?

For signalling to be applicable, there are some conditions that need to be met. First, if India has a good quality genetic pool then it should take costly steps to signal its quality. These steps will not be taken by countries that do not have a similarly promising set, or have a bad set, of genetic resources and ethnobotanical knowledge inputs. The assumption is that countries where genetic resources and ethnobotanical knowledge inputs are not of good quality, or are of a bad quality, will choose not to take steps to signal quality, since they would not generate sufficient profits to recoup the investments made in this regard, either in the mid-term or the long term.[11] In practice, this would mean that when access authorities embark upon the creation of inventories, if during the course of inventory generation it turns out that the genetic resources do not match the estimated value, then the access authority will discard the project since the additional costs cannot be recouped later on.

Both of these conditions are met in the market for bioprospecting. Since institutions for access and traditional knowledge impose large organizational costs on source countries, it makes sense to set them up only when the country has sufficient non-substitutable genetic resources (or demarcable communities with ethnobotanical knowledge) such that the benefits of the exercise can potentially offset the costs of such an effort. A country which does not meet this requirement of sufficient non-substitutable genetic resources or ethnobotanical knowledge knows that the costs it incurs in setting up such institutions would not be recouped via bioprospecting because even if it manages to attract investment for the first time from firms or intermediaries, they will not return.[12]

Hence, the setting up of effective access institutions and investment in sorting out the interface between rights on traditional knowledge and access within a country can very well serve as a signal of the quality of its genetic resources. Ten Kate and Wells (2001: 21) note similarly that,

"Biodiversity registers and integrated biodiversity databases are likely to attract partners, to support value-addition, to help protect intellectual property rights and thus to support a country's competitiveness."

Another way in which access authorities could signal the value of ethnobotanical knowledge (i.e. that they have high quality knowledge pools and not low quality ones) within their country is by offering collaborative contracts to bioprospectors. This is discussed in more detail below.

Screening

Firms need mechanisms that help them differentiate between holders of low quality and high quality knowledge pools. Similarly, access authorities would like to be able to distinguish the promising firms from their competitors when entering into bioprospecting collaborations. The solution lies in the creation of a so-called "separating equilibrium" through specific contractual offers.

In their landmark paper, Rothschild and Stiglitz (1976) have established the importance of separating equilibrium in screening mechanisms. By offering a particular type of contract, it is possible to separate the market into desirable and undesirable economic actors.[13]

A good way to create a separating equilibrium in bioprospecting would be for firms to offer contracts with benefit-sharing schemes that only attract holders of high quality knowledge pools. These would expressly refer to schemes wherein there are lower upfront payments and higher royalties. As a result of this only communities which are confident in their ability to add some value to the R&D process will come forward to accept such contracts.[14] In such a situation, a country with a low quality pool would not be able to recoup its investments (in setting up its institutions on CBD guidelines) when it has to accept low upfront payments and the eventual prospect of no profits from royalties due to the low success of the drug programmes.

Alternatively, access authorities can discriminate between firms through rules that ask for research details. National access laws can ask for information ranging from research goals, timeframes, to scientific capabilities, in order to assure themselves of the potential of the firm, its market stance, and its competitiveness.

The role of access in solving post-contractual moral hazard

In the post-contractual phase, the biggest concern for source countries is one of monitoring. In a majority of cases, it is extremely difficult to trace the flow of the genetic resources and associated ethnobotanical knowledge up the drug R&D chain and the proportional increase in its economic value. One way of preventing these *ex post* monitoring issues is to prefer upfront payments to royalty stipulations. But from a firm's perspective, incentive contracting with low upfront payments and higher royalties may be preferable to high upfront payments, since it reduces the risk of receiving traditional knowledge components and/or genetic resource supplies of low quality. Such contracts can also be an effective screening strategy for firms, as previously discussed.

Source countries can deal with *ex post* monitoring issues in two ways: reputation effects and a certification system.

Reputation effects

The evolution of the market for bioprospecting reveals that firms mainly prefer intermediaries in order to solve problems of adverse selection (see Chapter 4). There are a number of specialized intermediaries whose main activity is to collect genetic resources and/or ethnobotanical information to pass on to pharmaceutical firms. In doing this, these intermediaries transact *repeatedly* with the same national institutions. Such reputation effects can effectively solve the issue of opportunism by revealing the market potential of the collected resources more accurately, both to the source countries as well as to the firms. Thus, when intermediaries refrain from divulging the true market value of genetic resources, they can subsequently be refused access by the source countries. In the same way, when they misrepresent the value of genetic resources to firms, firms will not repeatedly transact with such intermediaries.

Source countries generally object to intermediaries on two grounds. First, when there are intermediaries, the task of monitoring end products based on their genetic resources and/or traditional knowledge components for benefit-sharing purposes is harder, since the access authorities cannot control contracts between intermediaries and pharmaceutical firms. Due to this, there is a general fear that their profit margins may be eroded. Secondly, source countries feel the pinch in terms of national implementation when it comes to harvesting and collecting genetic material. When there are too many intermediaries, the burden on access authorities for implementation is high.

Regional arrangements like the Declaration of Cancun adopted by the group of megadiverse countries in February 2002 can help them to use intermediaries and the concept of reputation effects to their advantage.[15] The Declaration contains an agreement amongst these countries to keep each other informed about cases of biopiracy that occur within their national frontiers. The firms and organizations that contribute to such biopiracy (including the end pharmaceutical concerns who buy such biopirated resources from intermediaries) are to be noted on a so-called "Black List", and those who figure on this "Black List" are to be denied access to all countries that form part of the megadiverse consortium

(Haas and Schwaegerl, 2002). Within legal literature, Coughlin (1993) is one of the few scholars who note the impact of reputation effects. He argues similarly that the failure to keep up with accepted obligations can result in a loss of access to other tropical countries due to the creation of mistrust.

Certification: The mandatory disclosure of PIC in patent applications

A certification system can be a useful way to create a conducive bargaining climate between users and providers by assuring source countries some control over the movement of genetic resources and/or traditional knowledge originating from within their territories. It may be especially helpful in the drug industry that is characterized by the presence of specialized intermediation whereby the ways and means in which genetic resources leave source countries are numerous. In such a scenario, a system of certification could act as proof, not only to the source countries that the genetic resources were obtained legally, but also to the end industry actors engaged in drug R&D, since it reduces legal uncertainty regarding the status of the R&D inputs (ten Kate and Laird, 1999). A certification system can also reduce the tendency amongst third parties to treat genetic resources as a public good. By requiring all users of genetic resources to comply with the certification requirement, it can act as an effective control over indiscriminate collection and harvesting activities within source countries. This is helpful in promoting biodiversity conservation.

A certification system can also help in several ways to clarify the interface with the patent system (Correa, 2004). For instance, it can help improve the substantive examination of patent applications involving biological materials and traditional knowledge, and also help in determination of prior art, thereby reducing the probability of bad patents being granted and in some cases, like drug research, even facilitate the invention (ibid.).

But there are several issues that need to be resolved before this can be considered an efficient option. One risk is that high transaction costs in obtaining such certificates caused by arbitrariness resulting from legal uncertainty or costly administrative procedures (in source countries) can hinder meaningful research or the use of genetic resources altogether (Glowka, 2001). Another disadvantage of such a system is that, since very few R&D programmes reach the stage where patents can be applied for in the case of pharmaceutical drugs and since a large share of the marketed products are not patent protected in the case of botanical medicines, requiring certificates to be a part of the patent application process would not effectively monitor compliance with the majority of benefit-sharing arrangements. Lastly, a certification system cannot be the main guarantor of benefit-sharing – it can only complement sound and effective laws on access and benefit-sharing (EC, 2001).

A certificate of source or a certificate of legal provenance have been also suggested as alternatives to the certificate of origin, in response to concerns that identification of the "origin" may prove impossible in many cases (see Cunningham et al., 2005, for a discussion).

The role of access in optimal risk sharing

It is true that shirking on the part of the communities or the national access institutions, or even other intermediaries, is a problem for firms, and that they prefer to rely on relatively small upfront payments and larger royalty payments to prevent a situation that may arise in the mid- or long term. For communities that survive in extremely hard and poor circumstances, waiting for, say, 15 years until a drug comes onto the market is not the best option. There are other possibilities through which both these competing interests could be solved.

Access laws can try and cater towards creating a better balancing of the risk attitudes of the parties in bioprospecting contracts. In particular, they can ensure that the risk-aversion of the communities is taken into account to a greater extent by such contracts, through options such as that of milestone payments.

Relation-specific investments in bioprospecting contracts and their impact on contractual behaviour

The last problem that still needs to be solved in the market for bioprospecting is that of asset specificity. The various actors in bioprospecting contracts, be they intermediaries, the communities, the access authorities, or the pharmaceutical firms, have their own physical and human assets.

Assuming that the national access authority performs its function and invests in building its institutional capabilities according to the CBD, detailed inventories of local genetic resources will be devised by such institutions. This is the first major asset of access authorities in bioprospecting contracts. For the communities, the main asset is their ethnobotanical knowledge. The ancestral lands on which they live are, in most national jurisdictions, considered to be their customary property but the right to alienate these lands may or may not be vested with the communities. Therefore, land cannot be counted as one of their alienable assets.

Assuming that ethnobotanical databases are compiled and that communities are represented by the access authority, the right to ethnobotanical knowledge of the uses contained in the database is the second asset of the access authority.

The main physical assets of the intermediaries are typical firm assets, such as some sort of laboratory equipment and human skills depending on their relative specializations. Intermediaries also have rights based on their inputs to the R&D process, such as the research results. The intermediaries also have their own reputation based on the quality of personnel and laboratory services and do not necessarily depend on one specific country for genetic resources. Although intermediaries have to contract for access with the access authority, they can specify the contract in more detail since they have restricted tasks to perform. They only need limited amounts of genetic resources, since they most likely do not have to take into account the possibility of clinical trials or drug development. Lastly, the results of research conducted at such intermediary stages by SMEs or DBFs can be traded to any larger firm on the spot market. Therefore, none of the assets of intermediaries are really relation-specific vis-à-vis the access authorities or the larger drug firms.

The pharmaceutical firms make one main relation-specific investment, which is their investment in R&D. The amounts of raw genetic resources required for R&D processes based on natural products are not negligible by any standards. More serious than ensuring the availability of such supplies during the research period, is the issue of how supply of the raw material is to be organized when a product based on it is discovered, which requires genetic material supplies for its production. This makes its R&D investments contingent on access contracts. Its investment is also contingent on legal certainty regarding the interface between its own intellectual property right and the *sui generis* intellectual property right on traditional knowledge.

When an access authority gets into a bioprospecting agreement, it employs both its assets – namely, the inventories as well as the ethnobotanical database. Asset specificity with regard to these two assets is created specifically because of intellectual property protection on final biotechnological inventions as under Article 27(3)(b) of the TRIPS Agreement. That is, once the pharmaceutical firm develops the final drug using these two assets, their first best use for both these inputs is the drug that the firm has developed. The second best use has no value because of intellectual property protection over the final product. Intellectual property protection prevents any other firm or agency from using the same genetic resource and traditional knowledge to derive the same product. The profits for the access authorities and the communities are therefore incumbent on the collaboration and continued relationship with the pharmaceutical firm.[16]

As a result, once the drug R&D process culminates in a marketable product for the production of which raw materials are required, the firm and the national access authority are very likely to be stuck in a bilateral monopoly. That is, the sole supplier of the raw materials for production (the national access authority) is faced by the sole demander (the production firm). Hence, the asset specificity of the firm is balanced by the counter dependence of the supplier. In a bilateral monopoly, both parties are in a bargaining situation but there is no simple rule to determine which one of the two will get a better bargain.[17]

At this stage, it is important to take a closer look at the relationship between the access authorities and the firm. The firm, when it enters into the bioprospecting contract, wants the assurance of reliable supplies. But it is impossible for the firm to specify *at the time of the contract* the exact quantity, quality, and precise time at which the raw materials are required, in contrast to, for example, a future market for agricultural products where the requirements are easily predictable. As a result, in the absence of an unconditional assurance of raw material supplies on behalf of the access authority at the start of the bioprospecting contract, the investment in research and development using genetic resources is always asset-specific for the firm.

So, *ex ante*, the firm would like to have access rights completely guaranteed at the start of the contract. But for the access authority, the problem is somewhat different. Since the relationship will turn into a bilateral monopoly when the developed use of the genetic resources or traditional knowledge is protected by intellectual property rights, the access authority will want to do this at a price which reflects a fair proportion of the future profit stream of the product. This price, according to the economic efficiency criteria of competitive markets, should be equal to the marginal productivity of the input (plant genetic resources and/or traditional knowledge) in the production process (including R&D) of the drug.

The task of the parties is to find an *ex ante* contractual structure which allocates the intellectual property right over ethnobotanical knowledge, intellectual property rights on the final drug product, and the right to access the particular genetic resource used in the product in such a way as to solve the problem of mutual hold-up.

The application of the property rights theory of the firm to bioprospecting demands that the control and ownership of these rights be allocated through contracts just as an ownership right of a more conventional asset, such as a machine. Economics of cumulative innovation proposes that intellectual property rights, once obtained in the form of patents or even trade secrets, be viewed as highly codified information. Accordingly, highly uncodified information becomes codified when products such as machines or other equipment have been created from it and the relevant information has been formalized through blueprints or patents on such products (Mandeville, 1999: 362). Viewed like this, intellectual property rights can be accounted for as assets of the firm in the drug industry.[18] Furthermore, it is possible to write contracts with intellectual property rights which separate contractible rights from residual rights (ownership). For example, a contract between a scientist franchising his patent to a manufacturing firm is a contract where the residual rights of control are still held by the scientist.

The right to access is a right vested with the national access authority. Access is not a strict ownership right, but more of a control right. Yet contracts can be written for access to all or some genetic resources of a country, or to one particular genetic resource within the country. These contracts also separate contractible rights from residual rights. That is, although the access authority grants the firm the right to "access" genetic resources, the residual control right is always vested in the access authority. *Incompleteness* in the GHM framework denotes that, at the *ex ante* stage, the exact form of the optimal transaction (that is, how many units, at what quality level, and the time of delivery, etc.) cannot be specified with certainty (Rogerson, 1992: 777, also see the discussion on the GHM framework in Chapter 4). This condition clearly applies to bioprospecting contracts. When a firm approaches a national access authority for an access contract, the firm cannot specify precisely how long it may need access in the first place. Secondly, it cannot specify which one of the genetic resources situated within the country it will need in large amounts for its research, nor at what time, and in what quantity.

As already discussed in Chapter 4, the assumptions of the property rights theory of the firm – namely, the definition of a firm as asset ownership, the incompleteness of contracts and the resulting specificity of investments – are all fulfilled in bioprospecting for drug R&D. An application of the results of this theoretical framework to the contractual problems in bioprospecting, as conducted in the next section, would therefore lead to useful recommendations and predictions for contractual structures between firms, intermediaries, communities, and access authorities.

Economic complementarity of assets in bioprospecting

Which of these property rights should be held together? Recapitulating the results from the discussion on asset specificity and property rights theory of the firm (see Chapter 4), two assets are complementary if they have positive effects on the marginal productivity of the agents only when they are jointly owned (Hart and Moore, 1990: 1135).

Applying this to bioprospecting, the final intellectual property right of firms on a drug, and the right to access that particular genetic resource needed for the production of the drug within the source country concerned, exhibit strong complementarities and should be jointly owned. To have the incentive to invest in R&D for a product that can be patented, the firm would like *ex ante* to have the residual control rights over access in case a product is discovered *ex post* whose production is contingent on the availability of the genetic resources. Therefore, the *undisturbed* right to access (that respects conservation aims) should be allocated to the firm through a contract *ex ante*.

It can also be argued that the right to ethnobotanical knowledge is a complementary asset to the firm's research in cases where it is a part of the R&D programme. Due to legal uncertainty caused by the definition of the right to traditional knowledge in national laws, the firm's rights on the product (and also the future of the R&D programme) is contingent on the condition that the community does not create any obstacle on its claims. Therefore, at the date of contract negotiation (date 0), in order to get back returns on its investment at the date of production (date 1), the firm would also like to determine the *ex ante* allocation of the right to ethnobotanical knowledge.

The suggestion of transfer of the right over ethnobotanical knowledge to the firm *ex ante* here is a suggestion for the transfer of only those components of the knowledge pool of the community that are contracted with the firm. The endowment of the right over ethnobotanical knowledge will still be with the community and their informed consent is also very important for contract formation. Respecting the community's rights over their resources and their informed consent in the contractual process is very important for distribution of benefits that arise from such commercialization. But efficiency of optimal R&D demands that at the time of entering the contract, at date 0, they contract away the right to those ethnobotanical components that they share in return for a specified benefit-sharing package, which could be anything from upfront payments to royalty rights, or both, as they may prefer.

In sum, economic logic would predict that *ex ante*, there should be a bundling of the rights of access, right to ethnobotanical knowledge, and the final intellectual property right. Such a property rights allocation would thereby be a more efficient one. The critical point that needs to be understood here is that the assets have to be held together for efficient specific investments to occur.

Does it matter who actually owns them? Should the firm own them or should the access authority or community be allowed to contract for the final intellectual property of the firm at the start of the R&D? The result of the property rights theory of the firm on indispensable agents suggests that an indispensable agent should be the owner of assets in order to maximize his or her incentives for specific investment. Furthermore, by making the indispensable agent the owner of assets, the incentives for *ex ante* investments remain unchanged for the other agents (see Chapter 4 for the definition of indispensable agent).

The indispensable agent in the case of bioprospecting is the pharmaceutical firm. The firm has all the other essential assets such as R&D facilities and intellectual property rights on other drugs and probably even production facilities. This would make the case for why the firm should be the contractual owner of the right to access and ethnobotanical knowledge from an *ex ante* perspective, as the firm is the only actor in the contractual relationship whose marginal productivity is enhanced by the joint ownership of the relevant property rights.

These results are very interesting for the bioprospecting market because legal scholars and national regulators have always argued for the reverse in the case of the right to access. The thrust is that the right to access should *never* be committed to fully *ex ante*, because it is the only major leverage for source countries to control sharing of benefits. This argument neglects the efficiency problems of specific investments completely.

It may be worth repeating here that the argument presented in this section does not prohibit the linking of the access right to benefit-sharing arrangements or other rules on how price for access to genetic resources should be set, or on limits to access to genetic resources based on conservation concerns. But, rather, that these rules for determining the price and conditions of access have to be transparent and defined in a comprehensive contract *ex ante* to ensure optimal R&D based on genetic resources.

Market response to concerns of asset specificity

If the scope of permissible contracts was unlimited, then the firms would ask for such an *ex ante* allocation of the rights to access and ethnobotanical knowledge, and joint ownership of the right to access and right to ethnobotanical knowledge, and final intellectual property rights on biotechnological inventions would be observed.

But the onus that the CBD places upon national access authorities to regulate externalities is one reason why the access right cannot be traded *ex ante* in this form, although from an industry perspective, this is most efficient. Given the general objectives of the CBD, the access authority cannot trade away the right to access without any restrictions on grounds of sustainable use and conservation, given the chance that the resource or resources from which the firm develops the drug in question may be abundantly available, but they may also be endangered or highly endangered.

Given the possibility that the access right might be truncated according to conservation needs of the genetic resource(s) in question, the invariable result is that although there is asset specificity, a long-term contract that allocates ownership rights *ex ante* is not an option. The drug industry responds to this in two main ways. The potential risk that the final product may be dependent on the availability of the natural product for its production shapes the R&D process. In the process of total synthesis of natural products, it is often possible to determine the essential features of the molecule necessary for activity (the pharmacophore). The desire to reduce potential dependency of the final drug on the availability of the natural product for its production can mean that researchers synthesize simpler analogues or synthetic "precursors" to the natural product that have similar or better activity (Cragg and Newman, 2004). A very good example of a synthetic "precursor" to the natural product which is now in clinical trials is the case of the marine-derived anti-cancer agent, halichondrin B (from the Japanese sponge *Halichondria okadai*). Halichondrin B and norhalichondrin B were synthesized in 1992, but synthetic schemes have been used since then to synthesize several variants of halicondrin B that are chemically more stable while maintaining the same level of biological activity (ibid.). Two of these agents have now been evaluated by the National Cancer Institute of the USA, in cooperation with Eisai Research Institute in the USA, and one of the compounds is in Phase 1 Clinical trials. Other examples where synthetic precursors have been developed for clinical trials are Dolostatin 10 (cytotoxic peptides isolated from the Indian Ocean mollusk, *Dolabella auricularia*, Bangamide derivatives (also isolated from the marine sponge) and HTI-286 (Hemiasterlin Derivative) (Newman and Cragg, 2004).

Industry also reveals a reliance on shorter contracts with the presence of intermediaries to deal with the problem of asset specific investments. In the bioprospecting market, reputation effects, coupled with competitive conditions for demand and supply, guide future interaction and prospects of efficient cooperation between specialized intermediaries although parties make substantial relation-specific investments. The market structure that can be seen in biotechnology-based research in general is therefore also characteristic of bioprospecting. The stages of collection of plant material, screening, and advanced screening, advanced research and development are all conducted with reputation and quality of cooperation as the benchmarks to ensure quality of research inputs at each stage and to deal with the problems of mutual trust that characterize this activity.

Notes

1. Normally, one could argue that a user fee or a tax denotes the content of the right (of the state) to price access and not just its form. But here, the content of rights are taken in the sense of details of how the right should actually be made operational, the capacity of the suggested rights structure to coordinate the kind of functions suggested and the institutional competence of the access authority.
2. See Pearce and Moran (1994), Polasky and Solow (1995), and McNeely (2000) among many others.
3. For example, Farnsworth and Soejarto (1985) predict that the foregone revenues due to plant extinction to the drug industry to be in excess of $1.5 million per species per year in 1980 dollars, whereas Mendelsohn and Balick (1995) estimate the net present value of as yet undiscovered drugs from rainforest plants to be $1.2 million per tropical plant species. The divergence in the studies is mainly due to the assumptions they are based on – whereas Farnsworth and Soejarto assume that 1 out of every 125 plants will yield a drug, Mendelsohn and Balick assume that only one out of a million plants tested will yield a new drug. For a detailed discussion, see Artuso (1997a: 4–6). Also see Artuso (1997b: 187–204). He notes that assumptions of R&D success rates, potential revenues of a new drug and the cost of capital can significantly affect the estimated value of screening biological material. This in turn can affect the final value that is allocated to finding a new drug from such screening activities (ibid.: 189).
4. As Artuso (1997a: 73–74) notes, the term "quasi-option value" seems to be in use to distinguish it from the option values literature in finance. See Arrow and Fiser (1974), Henry (1974), for some of the early applications of the option value literature to biodiversity issues. See Artuso (1997a: 73–74) for a discussion on the literature on option values.
5. Dhillon and Amundsen (2000: 114–115). They present a more elaborate list of species, out of which only a few selected have been reproduced here. Other scholars have also conducted exhaustive surveys of medicinal species that have grown to become extinct as a result of their medicinal value. See for example, Ayensu (1983).
6. The information contained in this paragraph is based on Sheldon et al. (1997: 20).
7. For example, the conservation agency of the government of Costa Rica, INBio, places a list of conditions for firms to be allowed to bioprospect. These are: high abundance, large number of specimens occurring within the inventory, worked-out taxonomy, easy access, easy to locate, occurring in protected areas. In addition to all this, the species' population should not be altered as a result of collection and it should preferably not be a herb (Dhillon and Amundsen, 2000: 109).
8. See generally, Brown (2000) for a discussion of "non-use" values of genetic resources. Also refer to Chapter 3.
9. Aylward (1996: 223) makes a good suggestion in this context. According to him, information that guides the selection of species for screening includes: collection information, taxonomical classification, ecological data regarding biochemical activity, ethnobotanical information about traditional uses of plants, among others, which the access authorities can make available for a fee in order to best capture the value of bioprospecting.
10. In this context, see Akerele (1990: 11) who notes: "Countries wishing to exploit fully their heritage of traditional medicine and the wealth of medicinal plants which most of them possess have a special interest in sponsoring ethnomedical studies, which bring together botanists, clinicians, pharmacologists and others. A logical first step, however, would be to review on a national basis the utilization of medical plants in general and the medicaments derived from them. Such an examination might reveal opportunities

for making greater use of safe and effective galenical preparations which might stimulate local cultivation and production and at the same time permit economies to be made."

11. Signalling works only if the signalling costs differ sufficiently amongst the "senders" of the signal. See Spence (1973, 1974).
12. See for example, Milgrom and Roberts (1986) for a formal game-theoretic version or Nelson (1974) for a non-formal version of the same concept. The India example presented here is similar to the argument used in these papers. These papers show or reason how (otherwise useless) advertising expenses can be used to signal product quality to consumers. It is based on the assumption that consumers would buy a high quality product again and again. A low quality product, in contrast, is useless. Once consumers know that it is useless, they will never buy it again. If a firm offers a high quality product it will use costly advertising. These expenses must be so large that they would not be made by a low quality firm so that people observing this expensive advertising will realize that it is a high quality firm. In other words, such advertising expenses can serve as a credible signal. Such a critical level of advertising exists because a high quality firm makes greater profits than a low quality one and enjoys the benefit of repeated purchases, and so can afford more advertising. This analysis assumes that consumers are able to objectively experience quality of goods. This advertisement model would not apply to markets with artificial product quality differentiation, like fashion goods.
13. In Akerlof's *market for lemons* example that has been discussed in Chapter 4, there is only one equilibrium and it is a pooling one. In contrast, a separating equilibrium can be created by contracts that only attract a particular kind of buyers. For example, insurance companies which want to separate high risk clients from low risk clients can offer contracts which fix an amount of personal liability for damages incurred. Such terms only attract low risk clients (and limit moral hazard). See Rothschild and Stiglitz (1976).
14. Lesser and Krattiger (1995) make the same point while noting the complexities of the contracting process vis-a-vis balancing the risks and the setting of the payment schemes.
15. The Like-Minded Megadiverse Group of Countries consists of Bolivia, Brazil, China, Colombia, Costa Rica, Ecuador, India, Indonesia, Kenya, Malaysia, Mexico, Peru, South Africa, and Venezuela.
16. The TRIPS Agreement grants product patent protection to pharmaceuticals as a result of which even if the communities and access authorities transact with the same genetic resource and ethnobotanical knowledge with other interested firms, since the already discovered use has intellectual property protection, these other firms will have to conduct extensive research once again until a final product may eventually be discovered that is different from the one on which intellectual property protection has already been granted.
17. Monopoly and monopsony will counteract each other in this situation. This does not mean that the market will end up with perfectly competitive prices, but in general, monopsony power will push price below marginal cost (the optimal price level) and monopoly power will push price closer to marginal revenue (which will be higher than the marginal cost). In bilateral monopolies, a single market equilibrium cannot be determined.
18. Hart (1995: 56) notes that the assets of the firm can consist of "hard" assets such as machinery, but they can also consist of "softer" assets such as patents, a firm's name or reputation.

7

Conclusions and recommendations

Developing countries have numerous issues to deal with in the area of drug research. Capacity building, affordable access to medicines, integrating traditional medicinal systems with modern health care, accountable drug procurement – are just some of them. Bioprospecting is only one solution to these problems; it is not a panacea. It is also not a measure that suits all – depending on the country in question, the genetic resources, and the kind of traditional knowledge available, the potential of the activity to bear fruitful results may vary.

Summary of results

This book has analysed the main issues in the design of national legal systems and institutions for bioprospecting. Such an analysis seems timely since the current uncertainty with regard to the nature and content of national laws on access and traditional knowledge has almost led to a freeze on bioprospecting for drug research, creating the need for legal policy action in countries that host genetic resources and traditional knowledge. The focus has been on the economic exchange process of actors in bioprospecting and the structure of the drug industry as the main basis for conducting the analysis. Using tools of law and economics, rights to access and traditional knowledge have been defined with the aim of promoting optimal contracting for drug R&D. The analysis has been conducted for both the pharmaceutical and botanical sectors, taking care to highlight the differences in intellectual property protection issues, traditional medicinal knowledge aspects and biodiversity conservation issues between the two sectors at every stage. The main results are as follows.

The legal analysis in Chapter 3 shows that a direct legal conflict between the provisions of the TRIPS Agreement and the CBD on the question of bioprospecting does not exist. Much of the controversy on Article 27(3)(b) of the TRIPS Agreement and Articles 8(j) and 15 of the CBD are a consequence of the vagueness of both these provisions and the contrasting political ambitions of countries as to how they perceive their individual gains. Countries need to recognize this and build consensus in the three spheres of interaction between intellectual property rights, biotechnology, and biodiversity *in order that* national bioprospecting frameworks can set out the rules and responsibilities of users and providers in access and traditional knowledge laws in an optimal way. It has also been shown in Chapter 3 that the right to access is the more important right for bioprospecting due to two main reasons. First, traditional knowledge is optional for firms that wish to bioprospect for drug research, but access to genetic resources is the *sine qua non* of such an endeavour. Second, upon a detailed analysis of the CBD, it is clear that the right to access is the best placed authority to implement many of its objectives with respect to sustainable use and conservation and to represent the communities.

An economic analysis of the market imperfections in the market for bioprospecting has been conducted in Chapter 4. The analysis lists the various transaction costs faced by parties to bioprospecting contracts that could be partially eliminated through well-defined property rights on access and traditional knowledge. It also shows that the kind of spot market transactions for genetic resources and traditional knowledge that one sees in bioprospecting is proof of low-level transactions. They lack complexity mainly because high-level transactions with complex contracting that are needed cannot emerge/be enforced in the market under existing constraints of ill-defined or non-existent property rights and insufficient legal regulation (Hart, 1995).

Successful attempts to define traditional knowledge should focus on demarcating the nature of the contribution that such knowledge could have rather than on the physical attributes of the right itself. The emphasis on the nature of the information itself serves as the best parameter of what the limits of "community/communities" are and what sort of knowledge ought to be protected and made contractible through an intellectual property right. Chapter 5 makes the case for defining a right over traditional medicinal knowledge that is representative of its contribution to drug R&D. Using the theory of cumulative innovation and evidence listed in investigations of ethnobotany and ethnopharmacology, it argues for a narrower definition for traditional knowledge, in the sense of ethnobotanical knowledge in the case of drug research. It shows that when traditional knowledge is defined in this narrower sense, two main incentive effects accrue: the incentive to keep the knowledge pool in its entirety and the incentive to reveal valuable information. Chapter 5 also shows that when communities are given a right to restrict access to outsiders to territories occupied by them in combination with the right to ethnobotanical knowledge, it can also encourage communities to conserve biodiversity at the site of its occurrence. Thus defined, the right can take the form of either a trade secret or a community intellectual property right. Some conditions that have to be met for both these forms of intellectual property protection have also been derived.

Such a rights definition that is based upon the contribution of knowledge to the R&D process, as well as clear-cut identification of beneficiaries who should enforce the right, is a precondition for benefits to be repatriated to the communities. This does not mean, however, that knowledge which falls outside the purview of ethnobotanical knowledge should not be protected. In fact, the thrust should be on conducting similar exercises in the case of traditional agricultural knowledge vis-à-vis agricultural biotechnology, traditional folklore vis-à-vis the music industry, and so on, so that well-contoured and enforceable rights can be derived.

But a well-defined right with an easily segregable set of beneficiaries is only the first step in empowering local and indigenous communities in this regard. A large onus rests on the design of institutions that will put this right into an enforceable framework. Whatever mode of intellectual property option is chosen for the right, the institutions have two major tasks: that of representing the communities effectively and of providing for rules of contract that take into account the difficulties of dealing with information as a resource. The checks and balances that should be in place to minimize principal–agent problems between communities (the principal) and the access authority (the agent) in contracts for ethnobotanical knowledge have been discussed in Chapter 6.

Many intriguing results flow out of the law and economic analysis of the access right. Chapter 6 has dealt with two very important questions: first, under what circumstances can regulation of access be justified for purposes of bioprospecting? Second, under what conditions is it worthwhile for the source country to set up costly institutional apparatus, as the CBD expects, in order to regulate access to genetic resources for bioprospecting?

Access regulation for bioprospecting can be justified only when externalities can be proven to result from drug R&D based on genetic resources. Chapter 6 establishes the clear-cut link between marketable products in the drug industry, their production process, and the unsustainable use of genetic resources that may occur in the pharmaceutical and botanical medicine sectors. It also establishes the conditions under which the potential revenues from bioprospecting for products in the drug industry can be high enough to offset the costs that source countries have to incur in setting up access institutions. Using results from environmental economics, it proposes individual transferable quotas as the appropriate rights structure for a right to access so that the biodiversity "hot spots" within source countries that hold extreme promise can be adequately protected through zoning and individual transferable quotas.

Appropriate redefinition of the rights to access and traditional knowledge in this way eradicates many of the market imperfections identified in Chapter 4. But those related to information asymmetries and uncertainty still remain. The second half of Chapter 6 shows that access institutions also play a critical role in providing contractual mechanisms to parties to deal with information asymmetries. Access institutions can help signal the quality of genetic resources and ethnobotanical knowledge within the country, and they can also screen for contract-worthy firms. It has been shown that when access institutions add value to in situ genetic diversity by way of creating inventories of genetic resources, and sort out the interface between ethnobotanical knowledge and modern drug research by investing in ethnobotanical databases, they can facilitate better bargaining conditions amongst parties.

The most important result of Chapter 6 relates to asset specific investments of parties in bioprospecting. If bioprospecting were to be one single strategic agreement, all parties would make asset specific investments. At the start of any R&D programme, it is uncertain whether any eventual drug that may be discovered will depend on the availability of a genetic resource for its production or whether its synthesis in an economically viable way will be possible. The drug firm, therefore, makes asset specific investments because its R&D, and ultimately its intellectual property right, is contingent on renewed access and legal certainty of traditional knowledge. The genetic resources of the source country and ethnobotanical components of the communities are also an asset specific investment because their first best use is in the drug which has been patented by the firm and the second best use is worthless. An application of the property rights theory of the firm reveals that all these three property rights have economic complementarities. For optimal incentives to invest in bioprospecting, there should be an *ex ante* bundling of these rights. Using the result on indispensable assets in property rights theory of the firm, it has been deduced that the firm should acquire these rights via contract *ex ante* to the R&D process.

Fairness and equity demand that initial endowment of property rights for access and ethnobotanical knowledge should be given to access authorities and communities respectively so that they can demand their share of benefits for their contributions. But for efficient bioprospecting, the right to access and the right over those ethnobotanical inputs that are revealed in the contract should be contracted away to the firm at the contract formation stage. The problem of monitoring profits to ensure that the firm then still shares profits with the communities and the access authorities can be enforced through mechanisms such as the international certification system or other contractual provisions.

In the absence of this ideal situation, industry response to the problem is twofold. First, when firms engage in natural products research, the emphasis is on deriving synthetic precursors of compounds with discovered medicinal uses so that their reliance on natural products for clinical trials and potentially large-scale production can be reduced. Second, the market relies on specialized intermediaries to supply genetic resources and/or traditional knowledge-related services because reputation effects, coupled with competitive conditions for demand and supply, support efficient cooperation in the presence of relation-specific investments.

Policy recommendations

Developing countries should focus on enacting effective laws and institutions for bioprospecting that take into account the realities of the drug R&D process. Although the laws themselves may not necessarily increase the transaction's legal certainty or certainty over the legal status of the materials transacted, an environment of uncertainty created by inadequate laws is a greater incentive for genetic resource users to go elsewhere or to choose other modes of R&D over natural products research (Glowka, 2001: 2). These laws should be enacted keeping in mind not only the rights of communities and access authorities in source countries, but also the demands of the process of drug R&D based on genetic resources.

Until now, strategies of source countries in the area of bioprospecting have been guided little by the fact that bioprospecting is only one potential solution to the public health and capacity-building issues that confront them. This needs to be revisited. Most issues faced by developing and least developed countries in the area of drug research are interlinked. The opportunity presented by bioprospecting should be viewed against this broader reality. If this is established, developing countries can, in return for access to genetic resources for established pharmaceutical firms, share in benefits in many other ways. One such benefit may be to bargain for the supply of drugs that established pharmaceutical firms hold patents on, such as those on malaria or some other tropical disease, so that the medicines can be obtained at cheaper rates, or even for free, in return for access to genetic resources.

Source country strategies also need to be guided by the fact that whereas developing countries have to be partners and liaise to negotiate for leg space at the national levels at international forums, in reality, when it comes to it, they are actually competitors in attracting investment into their genetic resources and/or traditional knowledge in order to harness their economic potential and capacity building through international collaborations, both private and public.

A staggering three billion people worldwide still rely on traditional medicines for their healthcare (Walsh, 2003), most of whom live in developing countries. Therefore, while attracting partners for bioprospecting collaborations, focus should be on optimizing the potential value of their genetic resources through expanding their collaborations with suitable screening organizations; by exposing their genetic resources to as many screens as possible, the chances of discovering a useful commercial drug are enhanced (Cragg, 2004, personal communication). A second focus should be on capacity building in such collaborative efforts, so that countries can bioprospect their resources themselves in order to sustain and enhance their traditional medicinal systems. Sustainable capacity building programmes should be initiated within developing countries, that are based on detailed surveys of their internal scientific and technological capacities. The gaps thus identified in internal scientific and technological capabilities should be sought to be filled up through international alliances and collaborations, including those in bioprospecting.

Whereas the negotiations for an international system of certification going on under the Conference of the Parties to the CBD are a very important step, PIC can only be one mechanism for ensuring benefit-sharing. Its limitations have to be adequately recognized and supplemented through contractual mechanisms. It should also be recognized that a system of certification can only help when firms apply for patents on their products. A vast majority of botanical medicines, for example, are protected as trade secrets – a system of certification will not be sufficient to deal with such products.

On the part of the firms, steps to enhance trust through the establishment of contractual best practices can help build mutually beneficial relationships.

Appendix I

Rights on access and traditional medicinal knowledge: A framework for implementation

The rights system for access: The implementation details

The regulation of access shall be the prerogative of the National Access Authority under the terms set out under the National Law of Access. The National Law of Access will constitute a National Access Authority at the centre. This National Access Authority is to be augmented in its role and responsibilities through regional offices for access at the state level. The National Access Authority shall have three main kinds of functions:

(a) Setting up a system of individual transferable quotas for the right to access based on the national biodiversity strategy and national action plan;
(b) Facilitating bioprospecting contracts;
(c) Monitoring and evaluation of contractual performance of firms and communities and of conservation and sustainable use activities within the country.

Antecedents of the rights system to access

The Access Authority's competence should be defined closely in coordination with national biodiversity strategies created in source countries in accordance with Article 6 of the CBD. If there is no national biodiversity strategy in place, the Access Authority should first embark upon the creation of a national biodiversity strategy. In this case, the Access

Authority will also be responsible for conducting the country study, analysing the descriptive data of the country study vis-à-vis the main pressures faced by the resources and the setting out of national priorities.

Based on the information gathered in the country study, the Access Law should direct the Access Authority to divide the flora and fauna in the country's special ecosystems into categories depending on their availability and resilience categories. Availability parameters – such as highly endangered, endangered, and abundantly available; and resilience parameters such as: can rejuvenate easily after exploitation, sensitive to exploitation, highly sensitive to exploitation – should be set out. Such areas should also be marked as protected areas, according to national strategies (see generally, Laird and Lisinge, 2002).

Based on this, the Access Authority has to define a national biodiversity action plan, which will contain:

(a) The action that is required to save the highly endangered and endangered species. This will mainly consist of the kinds of biodiversity programmes that need to be implemented in various parts of the country, and will need to be worked out in coordination with programmes that already exist within the country under other environmental protection and nature conservation laws;
(b) Decisions on how the National Access Authority will be complemented at the regional and local levels;
(c) The funding required to implement the programmes in the action plan successfully. Bioprospecting contracts should be used to fund the shortfall (at least partially) indicated by the difference between this estimate and the amount already available within the national budget.

Setting up an Individual Transferable Quota (ITQ) System for access to genetic resources

The Access Law will deem the ITQ system to be created only for those genetic resources that are found to be non-substitutable, based on the information in the country study.[1] This ITQ system for genetic resources will operate in a similar way to that for the fisheries sector of some of the developed countries, such as Norway, Australia, and New Zealand. The main purpose of the ITQs will be to create individual rights to a given share in the harvest of a surplus production of genetic resources. As in the case of fisheries, devising a successful ITQ system for genetic resources will require setting out parameters that take their spatial and inter-temporal dimensions into account. To respect the inter-temporal dimension (that is, the impact of present use on future availability), the organizational unit will need to be clearly set out. In the case of fisheries, this is based on fish population. But in the case of tropical forests, this might be the life cycle of the plants and trees involved. Setting out such organizational units is an exercise that has to be carried out in conjunction with scientific advice by the National Access Authority. To take into account the spatial dimension (that is, the impact of other species on valuable ecosystems), buffer zones should be created around these pristine ecosystems and the system of ITQs should be extended to these zones too.

Once the hot spots and buffer zones are demarcated in this way, the Access Authority will compute what can be called the *Total Allowable Harvest* (TAH).[2] This total allowable harvest will be based on the stock assessment, which will be computed using data of the various genetic resources collected under the country study – such as growth rates, mortality rates, fecundity by age class, etc. Under a system of ITQs, fixed quantities of the total allowable harvest will be allocated to firms and intermediaries or collectors, subject to periodic reallocation conditional on the availability of the genetic resources demanded by the bioprospectors.

The key feature of the ITQ system, as Symes and Crean (1995: 176) note, is that the quotas, once allocated may be freely traded (that is, sold or exchanged or leased) between users. This advantage is all the more important in the case of bioprospecting. Exchangeable and tradable quotas that can be passed on from less profitable to more profitable enterprises through market mechanisms to ensure that if the genetic resources have a high-end user value, this will get reflected in the price (ibid.: 176).

With the ITQ system in place, the Access Law will set out punitive fines and criminal liability (such as imprisonment) as punishment for anyone dealing with genetic resources without a valid ITQ. By making unauthorized dealings with genetic resources in contravention to the Access Law a criminal offence, this should act as a disincentive to people in source countries from dealing in genetic resources as though these were a public good.

The Access Law, through the ITQ system, will also take into account the interface between the right to access, which is only a negative right of the government and the property right of people on whose land the genetic resources may be situated. Once the Access Authority has decided on which lands its ITQ system will be applicable, based on the non-substitutability criteria, it will then deem that from thereon where areas are privately owned (for example, a patch of privately owned forest), owners can trade the property only through an ITQ system based on the constraints already set out by the Access Authority.

From a theoretical perspective, an ITQ system is perhaps the best property rights system to govern access to genetic resources. But in practice, such an ITQ system will depend on two factors for its success, namely, efficient institutional and enforcement mechanisms, and transferability, which is a key feature of the system. As McCay (1995: 2) notes, "ITQs are market-based management systems because with exclusive rights and transferability, markets can operate and take over from management agencies much of the role of allocating scarce resources." Therefore, the crux of the issue is: will markets really emerge for genetic resources? If not, the ITQ systems may get transformed into individual quota (IQ) systems, where transferability is highly restricted or not even possible.

If we take a closer look at the market for genetic resources and consider the way such an ITQ system for access will work, the following predictions can be made. Transferability of ITQs is high in the initial stages of pharmaceutical research – that is, when ITQs are distributed by the government for collection of any, say, 1,000 samples in rainforest area X, these samples can be traded easily. But as firms proceed to more advanced stages of research, such as clinical trials, and require huge amounts of a particular plant, it is quite unlikely that there will be many other firms interested in huge amounts of the same plant. For this reason, it is predictable that a market for tradable ITQs will not emerge at advanced stages of research. At these stages, the system will just reduce to being a quota system, where the role of the Access Authority will be to prescribe the conditions for use and the role of the firm will be to ensure that the harvest and collection will not exceed the quota prescribed.

Even apart from this limitation, high monitoring costs for the National Access Authority in deciding and allocating ITQs, coupled with transaction costs that could result from inefficient institutional set-ups and lack of proper enforcement, can lead to market failures. These need to be foreseen and prevented when ITQ systems are designed at national levels.

Making decisions on bioprospecting and facilitating the contracts

The Access Law will deem that bioprospecting will consist of two contracts in case the applicant is interested in access only: a contract for access and a contract for benefit-sharing. In cases where the applicant is interested in access and traditional medicinal knowledge, bioprospecting will consist of three different contracts: a contract for access, a contract for traditional medicinal knowledge, and a contract for benefit-sharing.

The access contract

The Access Law will specify that in order to qualify for a contract for access and obtain an ITQ the bioprospecting firm will have to agree to a list of several terms and conditions as set out by the Law. These are likely to include:

1. Complete revelation of research details.[3] This will be a disclosure clause which will oblige the applicant for access and/or traditional medicinal knowledge to disclose the following information:
   (a) The exact place where bioprospecting will be carried out and the components of the genetic resources (plants, animals, or micro-organisms) that will be used for research.
   (b) The goals of the research programme. These will consist of information on the kinds of medicines and research targets that the company has in mind, if it is the end pharmaceutical firm. Where the applicant is an intermediary organization, the information required here will concern the nature of the high-end firms and the research programmes that the collected genetic resources will be directed towards.
2. If the applicant is also interested in contracting for traditional medicinal knowledge, information about the kind of knowledge required and the name of the specific community or communities to be involved will need to be provided.
3. Acceptance of the condition to deposit voucher specimens and samples of all genetic materials collected within the country with the national authority.[4]
4. If the applicant is the end pharmaceutical firm, acceptance to deposit periodic reports on the progress of research. In cases where the applicant is an intermediary firm, an undertaking that the firm will inform the Access Authority about the recipient of the processed genetic resources for further research.
5. Acceptance that a separate agreement be made to the access contract for benefit-sharing, depending on whether the contract is for either genetic resources or traditional medicinal knowledge, or both.

If all these conditions are complied with, the bioprospector will be eligible for a contract for access with the Access Authority. In return for compliance to the above, the Access Law will list the following terms and conditions to be promised by the Access Authority to the firm in its Access Contract:

1. A declaration to maintain the research information deposited by the firm in complete confidentiality.
2. An *ex ante* declaration by the Access Authority that when the firm proceeds to the stage of clinical trials using any of the genetic resources within the country, it will be given continual access to the resources based only on the *Total Allowable Harvest* limitation – keeping in mind the sustainable use of the genetic resource in question.[5]
3. A clause on co-investment in research in selected cases.[6]

4. A clause stating that in cases where the Access Authority fails to keep within the terms of the agreement, the firm has judicial recourse in the source country against the actions of the Access Authority.

When the firm deposits all the information and is thus granted an Access Contract by the Access Authority, it will receive an ITQ along with the aforementioned assurances upon successful negotiation of the benefit-sharing contract (discussed below).

The contract for traditional medicinal knowledge

The Access Authority will represent the community whose medicinal knowledge is sought by the bioprospector in accordance with the terms agreed upon by them in the traditional medicinal knowledge deposition contract (see below the discussion on trade secret regime for traditional medicinal knowledge).

In addition to this, the contract will contain the general condition that when the Access Authority and the applicant conclude a contract for traditional medicinal knowledge with the consent of the communities, the Access Authority will give an *ex ante* declaration to the effect that the right to those components of traditional medicinal knowledge that will be marketed as trade secrets will be vested with the firm during the entire R&D process. Once products are developed using the trade secrets, the communities will not contest the patents on such products so long as the firm stays within the terms of benefit-sharing.

The firm will give an *ex ante* declaration to maintain the trade secrets during the course of research.

The benefit-sharing contract

The Access Law will contain different benefit-sharing options: such as milestone payments, technology transfer, upfront payments, royalty schemes. Using this scope for flexibility, the contracting parties (the firms or the intermediaries, the communities, and the Access Authority) will negotiate the benefit-sharing package that suits them best.

The nature of this benefit-sharing contract and the number of parties involved will depend on three things: (1) whether the applicant is interested only in access to genetic resources or whether it is also interested in traditional medicinal knowledge; (2) if interest is only in the former, whether application for access has been accepted and whether the Access Authority is ready to grant the ITQ; and (3) if interest is also shown in traditional medicinal knowledge, whether the community has expressed consent to accept such an offer to the Access Authority. If the answers are all positive, then the parties – the firm, the Access Authority, and representatives of the communities – will negotiate together on the terms of benefit-sharing.

To make these contracts transparent, the Access Law will set out the basis of setting payments for access and traditional medicinal knowledge. For payments for traditional medicinal knowledge, the Access Law will make a distinction between traditional medicinal knowledge that is used as a starting point of the drug R&D process and traditional medicinal knowledge that ends up contributing substantially to the final drug. It will provide that parties negotiate two different structures of payments *ex ante*, one of which will automatically be declared void at a later stage of the contract, depending on research results. This is very important to prevent *ex post* negotiations.

For payments in return for access, as already mentioned, the conservation programmes into which the money will be invested will be made clear by the Access Law. This will ensure transparency in the process. In a developing country context, this will also imply accountability where the failure of a government to invest the money generated therefrom into biodiversity programmes can be questioned by NGOs, researchers, and the citizens.

Monitoring

The Access Law will clearly provide for the Access Authority to have two kinds of monitoring functions: firstly, the monitoring and evaluation of contractual performance of firms and, secondly, the monitoring of conservation and sustainable use activities within the country.

Monitoring and evaluation of contractual performance of firms

The *ex ante* declarations on behalf of the Access Authority regarding the transfer of rights of access (through the ITQ system) and components of traditional medicinal knowledge contained in the Access Law solves the asset specificity problem of firms to a large extent. But the only way the asset specificity problem of the source country and the communities can be solved is through improved monitoring. Many standard clauses have been suggested in the preceding sections as part of the access contract and the traditional knowledge contract for monitoring purposes. In addition to all this, as a tool for monitoring, Access Laws should stipulate the issue of certificates of *Prior Informed Consent* as part of the contracts for access and traditional medicinal knowledge (latter to be issued by the communities).

In the case of alliances between countries, such as that developed by the mega-diverse countries in the Cancun declaration, where an international Black Book is maintained to record reputation effects, this provides a helpful monitoring tool, although outside the purview of the Access Law (see Chapter 6 for a discussion of this).

Monitoring of in situ *conservation activities*

The Access Law should also spell out that monitoring the implementation of the conservation programmes is a prerogative of the Access Authority and its regional offices. Only when efficient monitoring takes place can the Access Authority periodically update its categorization of flora and fauna based on availability and resilience factors. As we have already seen, it is very important for ITQs to be efficiently allocated.

In order to ensure biodiversity conservation in lands occupied by local and indigenous communities, these communities require rights over their physical territories, or what is called their "ancestral lands" (see, for example, Gehl Sampath and Tarasofsky, 2002). This is not only the right to usufruct (the right to use) but also the right to control access to their territories. For this, the communities have to be made the owners of the lands they possess so that they can restrict access of outsiders into their territories, which, as we have seen in Chapter 5, is the crux of *in situ* conservation. In this way, communities within these lands would be made the custodians of the genetic resources, subject to the general use restrictions placed by the Access Law.

A national Access Law also has to encourage communities to devise and implement their own *in situ* programmes with the help of revenues generated through bioprospecting. A good example of how communities can hold and implement their own in situ conservation programmes is shown by the case of the UZACHI group of indigenous people in Mexico (see Chapters 2 and 5).

Legal implementation of a trade secret regime for traditional medicinal knowledge

Devising and implementing a trade secret regime at the national level would consist of two different laws: one, a general trade secrecy law (or amendments to existing trade secrecy laws), and the other a law on traditional medicinal knowledge.

National trade secrecy law: The main components

A national trade secrecy law should be enacted using Article 39 of the TRIPS Agreement as its basis in order to give effective and stronger protection to trade secret holders, and in a way that is most suitable to cover ethnobotanical knowledge as a category of trade secrets. This would involve defining specific terms of relevance in a clear and consistent way. For example:

"*Legal Persons*": Article 39 only permits trade secrecy protection to legal persons, therefore, the national trade secrecy law should define "legal persons" to include groups of persons who share ethnobotanical knowledge and are registered as legal persons. An apparatus for legal registration should also be provided.

"*Reasonable Steps to Control the Information*": according to Article 39(3) of the TRIPS Agreement, in order for information to qualify for trade secret protection reasonable steps should have been taken by the person or persons in control of the knowledge to keep it secret. With this in mind, a national trade secrecy law should define "reasonable steps" to include rules for tribal organizations that exist within indigenous and local communities to control the use and dissemination of ethnobotanical knowledge.

"*Knowledge*" and "*Gross Negligence*": the definition of these terms is very important in order to protect holders of ethnobotanical knowledge from its possible misappropriation or sale to third parties by insiders – such as governments or private firms who document the knowledge as a trade secret, or even by the holders' own defecting group members. According to Article 39, third parties will be liable for illegal acquisition of trade secrets only if they have been grossly negligent in failing to know that dishonest commercial practices were used to acquire the information. The national trade secrecy law should define the concept of "knowledge" to imply a duty on the part of the buyer of the trade secret to obtain information about the possible incumbencies on the information (similar to the common law doctrine of notice that is prevalent in property transactions). When this is coupled with punishments of a punitive nature for trading without adequate notice and for theft of trade secrets, this will act as a deterrent for third parties from buying unauthorized trade secrets from defecting insiders.

Other details also need to be clarified by national trade secrecy laws. Some of the most important ones are that: trade secret protection is not time bound; cases for violation of trade secrets can be contested within any court in the country; and the theft of trade secrets is a punishable offence with penalties of a punitive nature.

National legislation on protection of traditional medicinal knowledge

The national law on traditional medicinal knowledge will deem that traditional medicinal knowledge is to be protected as trade secrets under the general terms defined under the national trade secret law. Towards this end, the legislation will provide that communities can be registered as legal persons and provide the detailed mechanism for the conversion of traditional medicinal knowledge as trade secrets.

Definition of traditional medicinal knowledge

The national traditional medicinal knowledge law will define traditional medicinal knowledge as knowledge of medicinal use of traditional and local communities or ethnobotanical knowledge, which can be registered and protected as a trade secret. It will define communities to mean those groups of people with a distinct body of medicinal knowledge held by written or oral systems (religious, social, or cultural), the members of which can be clearly demarcated and registered as 'legal entities' under the Act. This will ensure that only ethnobotanical knowledge qualifies as traditional medicinal knowledge under the law due to two different definitional aspects: firstly, that only knowledge held by segregable and demarcable communities that can come together to submit the information and be registered as "legal persons" will be covered by the law. Secondly, only knowledge which is not in the public domain will be protectable subject matter since, to qualify for trade secret protection, such information has to have been previously undisclosed.

Legal personality of tribes or communities

Tribes, communities, or even groups of tribes and communities, will approach the offices of the National Access Authority located within their state for registration as recognized "legal persons". The basic condition that needs to be fulfilled for them to be qualified as legal persons is that there should be a way to demarcate themselves (either one community or group, or sets of communities) from others around through rules of ancestry, tribal organization, lineage, ethnicity, or any other indicator. Only those groups that qualify to be registered as "legal persons" can initiate the process of documenting their knowledge with designated officers of the Office of the National Access Authority.

Protection of ethnobotanical knowledge as trade secrets

Ethnobotanical knowledge of such groups recognized as "legal persons" will then be catalogued in customized databases written in software such as FOXPRO software. An example of such a project currently being carried out is the conversion of traditional knowledge into trade secrets in Ecuador (Vogel, 1997). This is an exercise of validating the knowledge of such communities, and each regional office of the National Access Authority will act in conjunction with university researchers, scientists, and NGOs. The presence of these parties will ensure the scientific input needed in such a process, apart from lending objectivity to the principal–agent relationship between the communities and the National Access Authority. Each participating "legal entity" will have its own file in the database and will not be able to access the files of any other community in the database (ibid.).

Such ethnobotanical knowledge can be: (a) already in the public domain or (b) the same traditional knowledge can sometimes be held by different communities. To find out whether either of these two situations is true, the knowledge deposited by the communities is filtered by the regional access office against knowledge already held in the public domain worldwide through on-line botanical databases. For example, NAPRALERT software can be used to filter deposited traditional medicinal knowledge and to ascertain whether the knowledge is in the public domain or not. If the knowledge is not in the public domain, then it is registered as a trade secret in the file created for that community or set of communities. Such trade secrets can then be negotiated through a Material Transfer Agreement (MTA) with either the intermediaries or the final pharmaceutical firm by the National Access Authority. To find out whether many communities hold the same knowledge, the manager of the database filters the knowledge in the database across communities. If it is shown that two or more communities have the same knowledge, then the benefits that accrue from MTAs for the sharing of such knowledge have to be split equally among both or all the communities.

The traditional medicinal knowledge law will also provide for the terms and conditions of the relationship between communities that register their knowledge as trade secrets and the Offices of the Access Authority which register and represent them in bioprospecting contracts. This will be governed by what may be called the *Traditional Medicinal Knowledge Deposition Contract.*

The law will provide that communities which register as legal entities and start the process of conversion of their knowledge into trade secrets will sign a traditional medicinal knowledge deposition contract with the Office of the Access Authority, the provisions of which will be as follows.

The contract will contain an assurance by the agent, the National Access Authority, that the database will be safeguarded by the regional divisions of the National Access Authority through restricted entry to the database, apart from information protective measures and the practice of "due care", as is the case with company trade secrets (See Ganguli, 2000: 169 for a discussion of company practice). This will be similar to the confidentiality agreements that employees sign with their companies for protection of companies' trade secrets, the only difference here being that confidentiality will be promised by the Access Authority to the communities. This contract will also contain other terms of use of deposited secrets, such as: that the National Access Authority is allowed to enter into contracts only when authorized by the communities (with their prior informed consent); that the Access Authority will share with the communities all the relevant information the firm deposits on its research details to gain the right to access so that the communities can make informed decisions; that the National Access Authority will consult the communities at every stage of the contract (transparency); and that when any violation of the terms of this contract occurs, the communities can approach national courts. This contract will also contain a clause to the effect that representatives of the communities will be present when benefits are negotiated with the bioprospecting firm. These standard clauses, which will be assured to every community, are very important in order to reduce the principle–agent problems between communities and the national Access Authority. In addition to this, the law will also provide that communities are absolute owners of the land territories they occupy.

The project in Ecuador to convert traditional knowledge into trade secrets, discussed earlier, is a collaborative effort by the Inter American Development Bank–Consejo Nacional de Desarrollo Program on Environmental Capacity Building, CARE-Ecuador, and the NGO EcoCiencia. This is an independent project not supported by the government of Ecuador or by its national traditional knowledge law, but is evidence of the viability of this option and its large-scale potential (see Vogel, 1997, for more details).

National traditional medicinal knowledge laws will also provide that the knowledge which falls into the public domain will be catalogued and documented by the Access Authority and sent out to patent offices throughout the world to prevent wrongful patents being granted on them. Such traditional knowledge can be used in determining "prior art".

National traditional medicinal knowledge laws have also to lay down procedures through which communities can signify consent to the Access Authority in cases where there is a bioprospecting offer that they consider suitable. The best way to do this is for national laws to strengthen already existing decision-making mechanisms within communities or use them as the basis to devise consent procedures, instead of imposing wholly new sets of legal mechanisms (R.V. Anuradha, 2002, personal communication).

Thus, when there is an offer for bioprospecting from a pharmaceutical firm or an intermediary organization for the traditional medicinal knowledge of local and indigenous community X within a country, the application will be made to the Office of the Access Authority. Upon receiving such an offer, the Office of the Access Authority will act through its regional office and investigate whether the community has registered itself and deposited its traditional medicinal knowledge on its official database. If this is the case, the office will then advise community X of the possibility for the commercialization of the trade secrets through MTAs, together with other details on the research needs of the firm.[7] The community will use the procedure set out by the national traditional medicinal knowledge law to signify its prior informed consent. When the community decides it wishes to proceed with the offer, then the Access Authority will represent the community in the bioprospecting contract on terms of the traditional medicinal knowledge deposition contract that has been signed between community X and the Access Authority at the time when X deposited its traditional knowledge for conversion as trade secrets.

In effect, such an implementation mechanism will mean that when a bioprospector applies, the deposition contract between community and Access Authority is long in place, traditional medicinal knowledge has already been catalogued and is in the form of a trade secret, the knowledge which is in the public domain is also documented and left outside the purview of protection, and the rules of representation of the communities by the Access Authority are clearly set out by the traditional medicinal knowledge deposition contract. All this guarantees not only an expeditious process of getting a bioprospecting contract, it also ensures that different kinds of legal uncertainty do not harm the contract prospects *ex post.*

Notes

1. Since the country study is an inventory of genetic resources within a country, it will also shed light on which are the special ecosystems or "hot spots" and which are not. And as Chapter 6.1 has shown, the demand for bioprospecting will only be in those biodiversity "hot spots" where genetic resources are not substitutable with one another.
2. This concept of Total Allowable Harvest (TAH) is an adaptation of the concept of Total Allowable Catch (TAC) that is used in ITQs in the case of fisheries.
3. This information will enable the access authority eventually to screen the firms suitable for contracts for access and traditional knowledge.
4. It is important for the access authority to be able to monitor the firm or even, in extreme cases, contest it in court should the company indulge in post-contractual opportunism.
5. This clause will solve to a large extent the asset specificity problem of the firm.
6. This clause, when provided, will be useful for firms to screen interested national governments.
7. If it turns out that the community whose knowledge is in demand has not converted that knowledge into trade secrets using the machinery provided for it under the national traditional knowledge law, the access authority will, in the interest of the community, encourage it to do so.

Appendix II

Convention on biological diversity

5 JUNE 1992

Preamble

The Contracting Parties,

Conscious of the intrinsic value of biological diversity and of the ecological, genetic, social, economic, scientific, educational, cultural, recreational and aesthetic values of biological diversity and its components,

Conscious also of the importance of biological diversity for evolution and for maintaining life sustaining systems of the biosphere,

Affirming that the conservation of biological diversity is a common concern of humankind,

Reaffirming that States have sovereign rights over their own biological resources,

Reaffirming also that States are responsible for conserving their biological diversity and for using their biological resources in a sustainable manner,

Concerned that biological diversity is being significantly reduced by certain human activities,

Aware of the general lack of information and knowledge regarding biological diversity and of the urgent need to develop scientific, technical and institutional capacities to provide the basic understanding upon which to plan and implement appropriate measures,

Noting that it is vital to anticipate, prevent and attack the causes of significant reduction or loss of biological diversity at source,

Noting also that where there is a threat of significant reduction or loss of biological diversity, lack of full scientific certainty should not be used as a reason for postponing measures to avoid or minimize such a threat,

Noting further that the fundamental requirement for the conservation of biological diversity is the *in-situ* conservation of ecosystems and natural habitats and the maintenance and recovery of viable populations of species in their natural surroundings,

Noting further that *ex-situ* measures, preferably in the country of origin, also have an important role to play,

Recognizing the close and traditional dependence of many indigenous and local communities embodying traditional lifestyles on biological resources, and the desirability of sharing equitably benefits arising from the use of traditional knowledge, innovations and practices relevant to the conservation of biological diversity and the sustainable use of its components,

Recognizing also the vital role that women play in the conservation and sustainable use of biological diversity and affirming the need for the full participation of women at all levels of policy-making and implementation for biological diversity conservation,

Stressing the importance of, and the need to promote, international, regional and global cooperation among States and intergovernmental organizations and the non-governmental sector for the conservation of biological diversity and the sustainable use of its components,

Acknowledging that the provision of new and additional financial resources and appropriate access to relevant technologies can be expected to make a substantial difference in the world's ability to address the loss of biological diversity,

Acknowledging further that special provision is required to meet the needs of developing countries, including the provision of new and additional financial resources and appropriate access to relevant technologies,

Noting in this regard the special conditions of the least developed countries and small island States,

Acknowledging that substantial investments are required to conserve biological diversity and that there is the expectation of a broad range of environmental, economic and social benefits from those investments,

Recognizing that economic and social development and poverty eradication are the first and overriding priorities of developing countries,

Aware that conservation and sustainable use of biological diversity is of critical importance for meeting the food, health and other needs of the growing world population, for which purpose access to and sharing of both genetic resources and technologies are essential,

Noting that, ultimately, the conservation and sustainable use of biological diversity will strengthen friendly relations among States and contribute to peace for humankind,

Desiring to enhance and complement existing international arrangements for the conservation of biological diversity and sustainable use of its components, and

Determined to conserve and sustainably use biological diversity for the benefit of present and future generations,

Have agreed as follows:

Article 1. Objectives

The objectives of this Convention, to be pursued in accordance with its relevant provisions, are the conservation of biological diversity, the sustainable use of its components and the fair and equitable sharing of the benefits arising out of the utilization of genetic resources, including by appropriate access to genetic resources and by appropriate transfer of relevant technologies, taking into account all rights over those resources and to technologies, and by appropriate funding.

Article 2. Use of Terms

For the purposes of this Convention:

"*Biological diversity*" means the variability among living organisms from all sources including, *inter alia*, terrestrial, marine and other aquatic ecosystems and the ecological complexes of which they are part; this includes diversity within species, between species and of ecosystems.

"*Biological resources*" includes genetic resources, organisms or parts thereof, populations, or any other biotic component of ecosystems with actual or potential use or value for humanity.

"*Biotechnology*" means any technological application that uses biological systems, living organisms, or derivatives thereof, to make or modify products or processes for specific use.

"*Country of origin of genetic resources*" means the country which possesses those genetic resources in *in-situ* conditions.

"*Country providing genetic resources*" means the country supplying genetic resources collected from *in-situ* sources, including populations of both wild and domesticated species, or taken from *ex-situ* sources, which may or may not have originated in that country.

"*Domesticated or cultivated species*" means species in which the evolutionary process has been influenced by humans to meet their needs.

"*Ecosystem*" means a dynamic complex of plant, animal and micro-organism communities and their non-living environment interacting as a functional unit.

"*Ex-situ conservation*" means the conservation of components of biological diversity outside their natural habitats.

"*Genetic material*" means any material of plant, animal, microbial or other origin containing functional units of heredity.

"*Genetic resources*" means genetic material of actual or potential value.

"*Habitat*" means the place or type of site where an organism or population naturally occurs.

"*In-situ conditions*" means conditions where genetic resources exist within ecosystems and natural habitats, and, in the case of domesticated or cultivated species, in the surroundings where they have developed their distinctive properties.

"*In-situ conservation*" means the conservation of ecosystems and natural habitats and the maintenance and recovery of viable populations of species in their natural surroundings and, in the case of domesticated or cultivated species, in the surroundings where they have developed their distinctive properties.

"*Protected area*" means a geographically defined area which is designated or regulated and managed to achieve specific conservation objectives.

"*Regional economic integration organization*" means an organization constituted by sovereign States of a given region, to which its member States have transferred competence in respect of matters governed by this Convention and which has been duly authorized, in accordance with its internal procedures, to sign, ratify, accept, approve or accede to it.

"*Sustainable use*" means the use of components of biological diversity in a way and at a rate that does not lead to the long-term decline of biological diversity, thereby maintaining its potential to meet the needs and aspirations of present and future generations.

"*Technology*" includes biotechnology.

Article 3. Principle

States have, in accordance with the Charter of the United Nations and the principles of international law, the sovereign right to exploit their own resources pursuant to their own environmental policies, and the responsibility to ensure that activities within their jurisdiction or control do not cause damage to the environment of other States or of areas beyond the limits of national jurisdiction.

Article 4. Jurisdictional Scope

Subject to the rights of other States, and except as otherwise expressly provided in this Convention, the provisions of this Convention apply, in relation to each Contracting Party:

(a) In the case of components of biological diversity, in areas within the limits of its national jurisdiction; and
(b) In the case of processes and activities, regardless of where their effects occur, carried out under its jurisdiction or control, within the area of its national jurisdiction or beyond the limits of national jurisdiction.

Article 5. Cooperation

Each Contracting Party shall, as far as possible and as appropriate, cooperate with other Contracting Parties, directly or, where appropriate, through competent international organizations, in respect of areas beyond national jurisdiction and on other matters of mutual interest, for the conservation and sustainable use of biological diversity.

Article 6. General Measures for Conservation and Sustainable Use

Each Contracting Party shall, in accordance with its particular conditions and capabilities:

(a) Develop national strategies, plans or programmes for the conservation and sustainable use of biological diversity or adapt for this purpose existing strategies, plans or programmes which shall reflect, *inter alia*, the measures set out in this Convention relevant to the Contracting Party concerned; and
(b) Integrate, as far as possible and as appropriate, the conservation and sustainable use of biological diversity into relevant sectoral or cross-sectoral plans, programmes and policies.

Article 7. Identification and Monitoring

Each Contracting Party shall, as far as possible and as appropriate, in particular for the purposes of Articles 8 to 10:

(a) Identify components of biological diversity important for its conservation and sustainable use having regard to the indicative list of categories set down in Annex I;

(b) Monitor, through sampling and other techniques, the components of biological diversity identified pursuant to subparagraph (a) above, paying particular attention to those requiring urgent conservation measures and those which offer the greatest potential for sustainable use;

(c) Identify processes and categories of activities which have or are likely to have significant adverse impacts on the conservation and sustainable use of biological diversity, and monitor their effects through sampling and other techniques; and

(d) Maintain and organize, by any mechanism data, derived from identification and monitoring activities pursuant to subparagraphs (a), (b) and (c) above.

Article 8. In-situ Conservation

Each Contracting Party shall, as far as possible and as appropriate:

(a) Establish a system of protected areas or areas where special measures need to be taken to conserve biological diversity;

(b) Develop, where necessary, guidelines for the selection, establishment and management of protected areas or areas where special measures need to be taken to conserve biological diversity;

(c) Regulate or manage biological resources important for the conservation of biological diversity whether within or outside protected areas, with a view to ensuring their conservation and sustainable use;

(d) Promote the protection of ecosystems, natural habitats and the maintenance of viable populations of species in natural surroundings;

(e) Promote environmentally sound and sustainable development in areas adjacent to protected areas with a view to furthering protection of these areas;

(f) Rehabilitate and restore degraded ecosystems and promote the recovery of threatened species, *inter alia*, through the development and implementation of plans or other management strategies;

(g) Establish or maintain means to regulate, manage or control the risks associated with the use and release of living modified organisms resulting from biotechnology which are likely to have adverse environmental impacts that could affect the conservation and sustainable use of biological diversity, taking also into account the risks to human health;

(h) Prevent the introduction of, control or eradicate those alien species which threaten ecosystems, habitats or species;

(i) Endeavour to provide the conditions needed for compatibility between present uses and the conservation of biological diversity and the sustainable use of its components;

(j) Subject to its national legislation, respect, preserve and maintain knowledge, innovations and practices of indigenous and local communities embodying traditional lifestyles relevant for the conservation and sustainable use of biological diversity and promote their wider application with the approval and involvement of the holders of such knowledge, innovations and practices and encourage the equitable sharing of the benefits arising from the utilization of such knowledge, innovations and practices;

(k) Develop or maintain necessary legislation and/or other regulatory provisions for the protection of threatened species and populations;

(l) Where a significant adverse effect on biological diversity has been determined pursuant to Article 7, regulate or manage the relevant processes and categories of activities; and

(m) Cooperate in providing financial and other support for *in-situ* conservation outlined in subparagraphs (a) to (l) above, particularly to developing countries.

Article 9. Ex-situ Conservation

Each Contracting Party shall, as far as possible and as appropriate, and predominantly for the purpose of complementing *in-situ* measures:

(a) Adopt measures for the *ex-situ* conservation of components of biological diversity, preferably in the country of origin of such components;

(b) Establish and maintain facilities for *ex-situ* conservation of and research on plants, animals and micro-organisms, preferably in the country of origin of genetic resources;

(c) Adopt measures for the recovery and rehabilitation of threatened species and for their reintroduction into their natural habitats under appropriate conditions;

(d) Regulate and manage collection of biological resources from natural habitats for *ex-situ* conservation purposes so as not to threaten ecosystems and *in-situ* populations of species, except where special temporary *ex-situ* measures are required under subparagraph (c) above; and

(e) Cooperate in providing financial and other support for *ex-situ* conservation outlined in subparagraphs (a) to (d) above and in the establishment and maintenance of *ex-situ* conservation facilities in developing countries.

Article 10. Sustainable Use of Components of Biological Diversity

Each Contracting Party shall, as far as possible and as appropriate:

(a) Integrate consideration of the conservation and sustainable use of biological resources into national decision-making;

(b) Adopt measures relating to the use of biological resources to avoid or minimize adverse impacts on biological diversity;

(c) Protect and encourage customary use of biological resources in accordance with traditional cultural practices that are compatible with conservation or sustainable use requirements;

(d) Support local populations to develop and implement remedial action in degraded areas where biological diversity has been reduced; and

(e) Encourage cooperation between its governmental authorities and its private sector in developing methods for sustainable use of biological resources.

Article 11. Incentive Measures

Each Contracting Party shall, as far as possible and as appropriate, adopt economically and socially sound measures that act as incentives for the conservation and sustainable use of components of biological diversity.

Article 12. Research and Training

The Contracting Parties, taking into account the special needs of developing countries, shall:

(a) Establish and maintain programmes for scientific and technical education and training in measures for the identification, conservation and sustainable use of biological diversity and its components and provide support for such education and training for the specific needs of developing countries;

(b) Promote and encourage research which contributes to the conservation and sustainable use of biological diversity, particularly in developing countries, *inter alia*, in accordance with decisions of the Conference of the Parties taken in consequence of recommendations of the Subsidiary Body on Scientific, Technical and Technological Advice; and

(c) In keeping with the provisions of Articles 16, 18 and 20, promote and cooperate in the use of scientific advances in biological diversity research in developing methods for conservation and sustainable use of biological resources.

Article 13. Public Education and Awareness

The Contracting Parties shall:

(a) Promote and encourage understanding of the importance of, and the measures required for, the conservation of biological diversity, as well as its propagation through media, and the inclusion of these topics in educational programmes; and

(b) Cooperate, as appropriate, with other States and international organizations in developing educational and public awareness programmes, with respect to conservation and sustainable use of biological diversity.

Article 14. Impact Assessment and Minimizing Adverse Impacts

1. Each Contracting Party, as far as possible and as appropriate, shall:

(a) Introduce appropriate procedures requiring environmental impact assessment of its proposed projects that are likely to have significant adverse effects on biological diversity with a view to avoiding or minimizing such effects and, where appropriate, allow for public participation in such procedures;

(b) Introduce appropriate arrangements to ensure that the environmental consequences of its programmes and policies that are likely to have significant adverse impacts on biological diversity are duly taken into account;

(c) Promote, on the basis of reciprocity, notification, exchange of information and consultation on activities under their jurisdiction or control which are likely to significantly affect adversely the biological diversity of other States or areas beyond the limits of national jurisdiction, by encouraging the conclusion of bilateral, regional or multilateral arrangements, as appropriate;

(d) In the case of imminent or grave danger or damage, originating under its jurisdiction or control, to biological diversity within the area under jurisdiction of other States or in areas beyond the limits of national jurisdiction, notify immediately the potentially affected States of such danger or damage, as well as initiate action to prevent or minimize such danger or damage; and

(e) Promote national arrangements for emergency responses to activities or events, whether caused naturally or otherwise, which present a grave and imminent danger to biological diversity and encourage international cooperation to supplement such national efforts and, where appropriate and agreed by the States or regional economic integration organizations concerned, to establish joint contingency plans.

2. The Conference of the Parties shall examine, on the basis of studies to be carried out, the issue of liability and redress, including restoration and compensation, for damage to biological diversity, except where such liability is a purely internal matter.

Article 15. Access to Genetic Resources

1. Recognizing the sovereign rights of States over their natural resources, the authority to determine access to genetic resources rests with the national governments and is subject to national legislation.

2. Each Contracting Party shall endeavour to create conditions to facilitate access to genetic resources for environmentally sound uses by other Contracting Parties and not to impose restrictions that run counter to the objectives of this Convention.

3. For the purpose of this Convention, the genetic resources being provided by a Contracting Party, as referred to in this Article and Articles 16 and 19, are only those that are provided by Contracting Parties that are countries of origin of such resources or by the Parties that have acquired the genetic resources in accordance with this Convention.

4. Access, where granted, shall be on mutually agreed terms and subject to the provisions of this Article.

5. Access to genetic resources shall be subject to prior informed consent of the Contracting Party providing such resources, unless otherwise determined by that Party.

6. Each Contracting Party shall endeavour to develop and carry out scientific research based on genetic resources provided by other Contracting Parties with the full participation of, and where possible in, such Contracting Parties.

7. Each Contracting Party shall take legislative, administrative or policy measures, as appropriate, and in accordance with Articles 16 and 19 and, where necessary, through the financial mechanism established by Articles 20 and 21 with the aim of sharing in a fair and equitable way the results of research and development and the benefits arising from the commercial and other utilization of genetic resources with the Contracting Party providing such resources. Such sharing shall be upon mutually agreed terms.

Article 16. Access to and Transfer of Technology

1. Each Contracting Party, recognizing that technology includes biotechnology, and that both access to and transfer of technology among Contracting Parties are essential elements for the attainment of the objectives of this Convention, undertakes subject to the provisions of this Article to provide and/or facilitate access for and transfer to other Contracting Parties of technologies that are relevant to the conservation and sustainable use of biological diversity or make use of genetic resources and do not cause significant damage to the environment.

2. Access to and transfer of technology referred to in paragraph 1 above to developing countries shall be provided and/or facilitated under fair and most favourable terms, including on concessional and preferential terms where mutually agreed, and, where necessary, in accordance with the financial mechanism established by Articles 20 and 21. In the case of

technology subject to patents and other intellectual property rights, such access and transfer shall be provided on terms which recognize and are consistent with the adequate and effective protection of intellectual property rights. The application of this paragraph shall be consistent with paragraphs 3, 4 and 5 below.

3. Each Contracting Party shall take legislative, administrative or policy measures, as appropriate, with the aim that Contracting Parties, in particular those that are developing countries, which provide genetic resources are provided access to and transfer of technology which makes use of those resources, on mutually agreed terms, including technology protected by patents and other intellectual property rights, where necessary, through the provisions of Articles 20 and 21 and in accordance with international law and consistent with paragraphs 4 and 5 below.

4. Each Contracting Party shall take legislative, administrative or policy measures, as appropriate, with the aim that the private sector facilitates access to, joint development and transfer of technology referred to in paragraph 1 above for the benefit of both governmental institutions and the private sector of developing countries and in this regard shall abide by the obligations included in paragraphs 1, 2 and 3 above.

5. The Contracting Parties, recognizing that patents and other intellectual property rights may have an influence on the implementation of this Convention, shall cooperate in this regard subject to national legislation and international law in order to ensure that such rights are supportive of and do not run counter to its objectives.

Article 17. Exchange of Information

1. The Contracting Parties shall facilitate the exchange of information, from all publicly available sources, relevant to the conservation and sustainable use of biological diversity, taking into account the special needs of developing countries.

2. Such exchange of information shall include exchange of results of technical, scientific and socio-economic research, as well as information on training and surveying programmes, specialized knowledge, indigenous and traditional knowledge as such and in combination with the technologies referred to in Article 16, paragraph 1. It shall also, where feasible, include repatriation of information.

Article 18. Technical and Scientific Cooperation

1. The Contracting Parties shall promote international technical and scientific cooperation in the field of conservation and sustainable use of

biological diversity, where necessary, through the appropriate international and national institutions.

2. Each Contracting Party shall promote technical and scientific cooperation with other Contracting Parties, in particular developing countries, in implementing this Convention, *inter alia*, through the development and implementation of national policies. In promoting such cooperation, special attention should be given to the development and strengthening of national capabilities, by means of human resources development and institution building.

3. The Conference of the Parties, at its first meeting, shall determine how to establish a clearing-house mechanism to promote and facilitate technical and scientific cooperation.

4. The Contracting Parties shall, in accordance with national legislation and policies, encourage and develop methods of cooperation for the development and use of technologies, including indigenous and traditional technologies, in pursuance of the objectives of this Convention. For this purpose, the Contracting Parties shall also promote cooperation in the training of personnel and exchange of experts.

5. The Contracting Parties shall, subject to mutual agreement, promote the establishment of joint research programmes and joint ventures for the development of technologies relevant to the objectives of this Convention.

Article 19. Handling of Biotechnology and Distribution of its Benefits

1. Each Contracting Party shall take legislative, administrative or policy measures, as appropriate, to provide for the effective participation in biotechnological research activities by those Contracting Parties, especially developing countries, which provide the genetic resources for such research, and where feasible in such Contracting Parties.

2. Each Contracting Party shall take all practicable measures to promote and advance priority access on a fair and equitable basis by Contracting Parties, especially developing countries, to the results and benefits arising from biotechnologies based upon genetic resources provided by those Contracting Parties. Such access shall be on mutually agreed terms.

3. The Parties shall consider the need for and modalities of a protocol setting out appropriate procedures, including, in particular, advance informed agreement, in the field of the safe transfer, handling and use of any living modified organism resulting from biotechnology that may have adverse effect on the conservation and sustainable use of biological diversity.

4. Each Contracting Party shall, directly or by requiring any natural or legal person under its jurisdiction providing the organisms referred to in paragraph 3 above, provide any available information about the use and safety regulations required by that Contracting Party in handling such organisms, as well as any available information on the potential adverse impact of the specific organisms concerned to the Contracting Party into which those organisms are to be introduced.

Article 20. Financial Resources

1. Each Contracting Party undertakes to provide, in accordance with its capabilities, financial support and incentives in respect of those national activities which are intended to achieve the objectives of this Convention, in accordance with its national plans, priorities and programmes.

2. The developed country Parties shall provide new and additional financial resources to enable developing country Parties to meet the agreed full incremental costs to them of implementing measures which fulfil the obligations of this Convention and to benefit from its provisions and which costs are agreed between a developing country Party and the institutional structure referred to in Article 21, in accordance with policy, strategy, programme priorities and eligibility criteria and an indicative list of incremental costs established by the Conference of the Parties. Other Parties, including countries undergoing the process of transition to a market economy, may voluntarily assume the obligations of the developed country Parties. For the purpose of this Article, the Conference of the Parties, shall at its first meeting establish a list of developed country Parties and other Parties which voluntarily assume the obligations of the developed country Parties. The Conference of the Parties shall periodically review and if necessary amend the list. Contributions from other countries and sources on a voluntary basis would also be encouraged. The implementation of these commitments shall take into account the need for adequacy, predictability and timely flow of funds and the importance of burden-sharing among the contributing Parties included in the list.

3. The developed country Parties may also provide, and developing country Parties avail themselves of, financial resources related to the implementation of this Convention through bilateral, regional and other multilateral channels.

4. The extent to which developing country Parties will effectively implement their commitments under this Convention will depend on the effective implementation by developed country Parties of their commitments under this Convention related to financial resources and transfer of tech-

nology and will take fully into account the fact that economic and social development and eradication of poverty are the first and overriding priorities of the developing country Parties.

5. The Parties shall take full account of the specific needs and special situation of least developed countries in their actions with regard to funding and transfer of technology.

6. The Contracting Parties shall also take into consideration the special conditions resulting from the dependence on, distribution and location of, biological diversity within developing country Parties, in particular small island States.

7. Consideration shall also be given to the special situation of developing countries, including those that are most environmentally vulnerable, such as those with arid and semi-arid zones, coastal and mountainous areas.

Article 21. Financial Mechanism

1. There shall be a mechanism for the provision of financial resources to developing country Parties for purposes of this Convention on a grant or concessional basis the essential elements of which are described in this Article. The mechanism shall function under the authority and guidance of, and be accountable to, the Conference of the Parties for purposes of this Convention. The operations of the mechanism shall be carried out by such institutional structure as may be decided upon by the Conference of the Parties at its first meeting. For purposes of this Convention, the Conference of the Parties shall determine the policy, strategy, programme priorities and eligibility criteria relating to the access to and utilization of such resources. The contributions shall be such as to take into account the need for predictability, adequacy and timely flow of funds referred to in Article 20 in accordance with the amount of resources needed to be decided periodically by the Conference of the Parties and the importance of burden-sharing among the contributing Parties included in the list referred to in Article 20, paragraph 2. Voluntary contributions may also be made by the developed country Parties and by other countries and sources. The mechanism shall operate within a democratic and transparent system of governance.

2. Pursuant to the objectives of this Convention, the Conference of the Parties shall at its first meeting determine the policy, strategy and programme priorities, as well as detailed criteria and guidelines for eligibility for access to and utilization of the financial resources including monitoring and evaluation on a regular basis of such utilization. The Conference of the Parties shall decide on the arrangements to give effect to para-

graph 1 above after consultation with the institutional structure entrusted with the operation of the financial mechanism.

3. The Conference of the Parties shall review the effectiveness of the mechanism established under this Article, including the criteria and guidelines referred to in paragraph 2 above, not less than two years after the entry into force of this Convention and thereafter on a regular basis. Based on such review, it shall take appropriate action to improve the effectiveness of the mechanism if necessary.

4. The Contracting Parties shall consider strengthening existing financial institutions to provide financial resources for the conservation and sustainable use of biological diversity.

Article 22. Relationship with Other International Conventions

1. The provisions of this Convention shall not affect the rights and obligations of any Contracting Party deriving from any existing international agreement, except where the exercise of those rights and obligations would cause a serious damage or threat to biological diversity.

2. Contracting Parties shall implement this Convention with respect to the marine environment consistently with the rights and obligations of States under the law of the sea.

Article 23. Conference of the Parties

1. A Conference of the Parties is hereby established. The first meeting of the Conference of the Parties shall be convened by the Executive Director of the United Nations Environment Programme not later than one year after the entry into force of this Convention. Thereafter, ordinary meetings of the Conference of the Parties shall be held at regular intervals to be determined by the Conference at its first meeting.

2. Extraordinary meetings of the Conference of the Parties shall be held at such other times as may be deemed necessary by the Conference, or at the written request of any Party, provided that, within six months of the request being communicated to them by the Secretariat, it is supported by at least one third of the Parties.

3. The Conference of the Parties shall by consensus agree upon and adopt rules of procedure for itself and for any subsidiary body it may establish, as well as financial rules governing the funding of the Secretariat. At each ordinary meeting, it shall adopt a budget for the financial period until the next ordinary meeting.

4. The Conference of the Parties shall keep under review the implementation of this Convention, and, for this purpose, shall:

(a) Establish the form and the intervals for transmitting the information to be submitted in accordance with Article 26 and consider such information as well as reports submitted by any subsidiary body;

(b) Review scientific, technical and technological advice on biological diversity provided in accordance with Article 25;

(c) Consider and adopt, as required, protocols in accordance with Article 28;

(d) Consider and adopt, as required, in accordance with Articles 29 and 30, amendments to this Convention and its annexes;

(e) Consider amendments to any protocol, as well as to any annexes thereto, and, if so decided, recommend their adoption to the parties to the protocol concerned;

(f) Consider and adopt, as required, in accordance with Article 30, additional annexes to this Convention;

(g) Establish such subsidiary bodies, particularly to provide scientific and technical advice, as are deemed necessary for the implementation of this Convention;

(h) Contact, through the Secretariat, the executive bodies of conventions dealing with matters covered by this Convention with a view to establishing appropriate forms of cooperation with them; and

(i) Consider and undertake any additional action that may be required for the achievement of the purposes of this Convention in the light of experience gained in its operation.

5. The United Nations, its specialized agencies and the International Atomic Energy Agency, as well as any State not Party to this Convention, may be represented as observers at meetings of the Conference of the Parties. Any other body or agency, whether governmental or non-governmental, qualified in fields relating to conservation and sustainable use of biological diversity, which has informed the Secretariat of its wish to be represented as an observer at a meeting of the Conference of the Parties, may be admitted unless at least one third of the Parties present object. The admission and participation of observers shall be subject to the rules of procedure adopted by the Conference of the Parties.

Article 24. Secretariat

1. A secretariat is hereby established. Its functions shall be:

(a) To arrange for and service meetings of the Conference of the Parties provided for in Article 23;

(b) To perform the functions assigned to it by any protocol;

(c) To prepare reports on the execution of its functions under this Convention and present them to the Conference of the Parties;

(d) To coordinate with other relevant international bodies and, in particular to enter into such administrative and contractual arrangements as may be required for the effective discharge of its functions; and

(e) To perform such other functions as may be determined by the Conference of the Parties.

2. At its first ordinary meeting, the Conference of the Parties shall designate the secretariat from amongst those existing competent international organizations which have signified their willingness to carry out the secretariat functions under this Convention.

Article 25. Subsidiary Body on Scientific, Technical and Technological Advice

1. A subsidiary body for the provision of scientific, technical and technological advice is hereby established to provide the Conference of the Parties and, as appropriate, its other subsidiary bodies with timely advice relating to the implementation of this Convention. This body shall be open to participation by all Parties and shall be multidisciplinary. It shall comprise government representatives competent in the relevant field of expertise. It shall report regularly to the Conference of the Parties on all aspects of its work.

2. Under the authority of and in accordance with guidelines laid down by the Conference of the Parties, and upon its request, this body shall:

(a) Provide scientific and technical assessments of the status of biological diversity;

(b) Prepare scientific and technical assessments of the effects of types of measures taken in accordance with the provisions of this Convention;

(c) Identify innovative, efficient and state-of-the-art technologies and know-how relating to the conservation and sustainable use of biological diversity and advise on the ways and means of promoting development and/or transferring such technologies;

(d) Provide advice on scientific programmes and international cooperation in research and development related to conservation and sustainable use of biological diversity; and

(e) Respond to scientific, technical, technological and methodological questions that the Conference of the Parties and its subsidiary bodies may put to the body.

3. The functions, terms of reference, organization and operation of this body may be further elaborated by the Conference of the Parties.

Article 26. Reports

Each Contracting Party shall, at intervals to be determined by the Conference of the Parties, present to the Conference of the Parties, reports on measures which it has taken for the implementation of the provisions of this Convention and their effectiveness in meeting the objectives of this Convention.

Article 27. Settlement of Disputes

1. In the event of a dispute between Contracting Parties concerning the interpretation or application of this Convention, the parties concerned shall seek solution by negotiation.

2. If the parties concerned cannot reach agreement by negotiation, they may jointly seek the good offices of, or request mediation by, a third party.

3. When ratifying, accepting, approving or acceding to this Convention, or at any time thereafter, a State or regional economic integration organization may declare in writing to the Depositary that for a dispute not resolved in accordance with paragraph 1 or paragraph 2 above, it accepts one or both of the following means of dispute settlement as compulsory:

(a) Arbitration in accordance with the procedure laid down in Part 1 of Annex II;

(b) Submission of the dispute to the International Court of Justice.

4. If the parties to the dispute have not, in accordance with paragraph 3 above, accepted the same or any procedure, the dispute shall be submitted to conciliation in accordance with Part 2 of Annex II unless the parties otherwise agree.

5. The provisions of this Article shall apply with respect to any protocol except as otherwise provided in the protocol concerned.

Article 28. Adoption of Protocols

1. The Contracting Parties shall cooperate in the formulation and adoption of protocols to this Convention.

2. Protocols shall be adopted at a meeting of the Conference of the Parties.

3. The text of any proposed protocol shall be communicated to the Contracting Parties by the Secretariat at least six months before such a meeting.

Article 29. Amendment of the Convention or Protocols

1. Amendments to this Convention may be proposed by any Contracting Party. Amendments to any protocol may be proposed by any Party to that protocol.

2. Amendments to this Convention shall be adopted at a meeting of the Conference of the Parties. Amendments to any protocol shall be adopted at a meeting of the Parties to the Protocol in question. The text of any proposed amendment to this Convention or to any protocol, except as may otherwise be provided in such protocol, shall be communicated to the Parties to the instrument in question by the secretariat at least six months before the meeting at which it is proposed for adoption. The secretariat shall also communicate proposed amendments to the signatories to this Convention for information.

3. The Parties shall make every effort to reach agreement on any proposed amendment to this Convention or to any protocol by consensus. If all efforts at consensus have been exhausted, and no agreement reached, the amendment shall as a last resort be adopted by a two-third majority vote of the Parties to the instrument in question present and voting at the meeting, and shall be submitted by the Depositary to all Parties for ratification, acceptance or approval.

4. Ratification, acceptance or approval of amendments shall be notified to the Depositary in writing. Amendments adopted in accordance with paragraph 3 above shall enter into force among Parties having accepted them on the ninetieth day after the deposit of instruments of ratification, acceptance or approval by at least two thirds of the Contracting Parties to this Convention or of the Parties to the protocol concerned, except as may otherwise be provided in such protocol. Thereafter the amendments shall enter into force for any other Party on the ninetieth day after that Party deposits its instrument of ratification, acceptance or approval of the amendments.

5. For the purposes of this Article, "Parties present and voting" means Parties present and casting an affirmative or negative vote.

Article 30. Adoption and Amendment of Annexes

1. The annexes to this Convention or to any protocol shall form an integral part of the Convention or of such protocol, as the case may be, and, unless expressly provided otherwise, a reference to this Convention or its protocols constitutes at the same time a reference to any annexes thereto. Such annexes shall be restricted to procedural, scientific, technical and administrative matters.

2. Except as may be otherwise provided in any protocol with respect to its annexes, the following procedure shall apply to the proposal, adoption and entry into force of additional annexes to this Convention or of annexes to any protocol:

(a) Annexes to this Convention or to any protocol shall be proposed and adopted according to the procedure laid down in Article 29;

(b) Any Party that is unable to approve an additional annex to this Convention or an annex to any protocol to which it is Party shall so notify the Depositary, in writing, within one year from the date of the communication of the adoption by the Depositary. The Depositary shall without delay notify all Parties of any such notification received. A Party may at any time withdraw a previous declaration of objection and the annexes shall thereupon enter into force for that Party subject to subparagraph (c) below;

(c) On the expiry of one year from the date of the communication of the adoption by the Depositary, the annex shall enter into force for all Parties to this Convention or to any protocol concerned which have not submitted a notification in accordance with the provisions of subparagraph (b) above.

3. The proposal, adoption and entry into force of amendments to annexes to this Convention or to any protocol shall be subject to the same procedure as for the proposal, adoption and entry into force of annexes to the Convention or annexes to any protocol.

4. If an additional annex or an amendment to an annex is related to an amendment to this Convention or to any protocol, the additional annex or amendment shall not enter into force until such time as the amendment to the Convention or to the protocol concerned enters into force.

Article 31. Right to Vote

1. Except as provided for in paragraph 2 below, each Contracting Party to this Convention or to any protocol shall have one vote.

2. Regional economic integration organizations, in matters within their competence, shall exercise their right to vote with a number of votes equal to the number of their member States which are Contracting Parties to this Convention or the relevant protocol. Such organizations shall not exercise their right to vote if their member States exercise theirs, and vice versa.

Article 32. Relationship between this Convention and Its Protocols

1. A State or a regional economic integration organization may not become a Party to a protocol unless it is, or becomes at the same time, a Contracting Party to this Convention.

2. Decisions under any protocol shall be taken only by the Parties to the

protocol concerned. Any Contracting Party that has not ratified, accepted or approved a protocol may participate as an observer in any meeting of the parties to that protocol.

Article 33. Signature

This Convention shall be open for signature at Rio de Janeiro by all States and any regional economic integration organization from 5 June 1992 until 14 June 1992, and at the United Nations Headquarters in New York from 15 June 1992 to 4 June 1993.

Article 34. Ratification, Acceptance or Approval

1. This Convention and any protocol shall be subject to ratification, acceptance or approval by States and by regional economic integration organizations. Instruments of ratification, acceptance or approval shall be deposited with the Depositary.

2. Any organization referred to in paragraph 1 above which becomes a Contracting Party to this Convention or any protocol without any of its member States being a Contracting Party shall be bound by all the obligations under the Convention or the protocol, as the case may be. In the case of such organizations, one or more of whose member States is a Contracting Party to this Convention or relevant protocol, the organization and its member States shall decide on their respective responsibilities for the performance of their obligations under the Convention or protocol, as the case may be. In such cases, the organization and the member States shall not be entitled to exercise rights under the Convention or relevant protocol concurrently.

3. In their instruments of ratification, acceptance or approval, the organizations referred to in paragraph 1 above shall declare the extent of their competence with respect to the matters governed by the Convention or the relevant protocol. These organizations shall also inform the Depositary of any relevant modification in the extent of their competence.

Article 35. Accession

1. This Convention and any protocol shall be open for accession by States and by regional economic integration organizations from the date on which the Convention or the protocol concerned is closed for signature. The instruments of accession shall be deposited with the Depositary.

2. In their instruments of accession, the organizations referred to in

paragraph 1 above shall declare the extent of their competence with respect to the matters governed by the Convention or the relevant protocol. These organizations shall also inform the Depositary of any relevant modification in the extent of their competence.

3. The provisions of Article 34, paragraph 2, shall apply to regional economic integration organizations which accede to this Convention or any protocol.

Article 36. Entry Into Force

1. This Convention shall enter into force on the ninetieth day after the date of deposit of the thirtieth instrument of ratification, acceptance, approval or accession.

2. Any protocol shall enter into force on the ninetieth day after the date of deposit of the number of instruments of ratification, acceptance, approval or accession, specified in that protocol, has been deposited.

3. For each Contracting Party which ratifies, accepts or approves this Convention or accedes thereto after the deposit of the thirtieth instrument of ratification, acceptance, approval or accession, it shall enter into force on the ninetieth day after the date of deposit by such Contracting Party of its instrument of ratification, acceptance, approval or accession.

4. Any protocol, except as otherwise provided in such protocol, shall enter into force for a Contracting Party that ratifies, accepts or approves that protocol or accedes thereto after its entry into force pursuant to paragraph 2 above, on the ninetieth day after the date on which that Contracting Party deposits its instrument of ratification, acceptance, approval or accession, or on the date on which this Convention enters into force for that Contracting Party, whichever shall be the later.

5. For the purposes of paragraphs 1 and 2 above, any instrument deposited by a regional economic integration organization shall not be counted as additional to those deposited by member States of such organization.

Article 37. Reservations

No reservations may be made to this Convention.

Article 38. Withdrawals

1. At any time after two years from the date on which this Convention has entered into force for a Contracting Party, that Contracting Party may withdraw from the Convention by giving written notification to the Depositary.

2. Any such withdrawal shall take place upon expiry of one year after

the date of its receipt by the Depositary, or on such later date as may be specified in the notification of the withdrawal.

3. Any Contracting Party which withdraws from this Convention shall be considered as also having withdrawn from any protocol to which it is party.

Article 39. Financial Interim Arrangements

Provided that it has been fully restructured in accordance with the requirements of Article 21, the Global Environment Facility of the United Nations Development Programme, the United Nations Environment Programme and the International Bank for Reconstruction and Development shall be the institutional structure referred to in Article 21 on an interim basis, for the period between the entry into force of this Convention and the first meeting of the Conference of the Parties or until the Conference of the Parties decides which institutional structure will be designated in accordance with Article 21.

Article 40. Secretariat Interim Arrangements

The secretariat to be provided by the Executive Director of the United Nations Environment Programme shall be the secretariat referred to in Article 24, paragraph 2, on an interim basis for the period between the entry into force of this Convention and the first meeting of the Conference of the Parties.

Article 41. Depositary

The Secretary-General of the United Nations shall assume the functions of Depositary of this Convention and any protocols.

Article 42. Authentic Texts

The original of this Convention, of which the Arabic, Chinese, English, French, Russian and Spanish texts are equally authentic, shall be deposited with the Secretary-General of the United Nations.

IN WITNESS WHEREOF the undersigned, being duly authorized to that effect, have signed this Convention.

Done at Rio de Janeiro on this fifth day of June, one thousand nine hundred and ninety-two.

Annex I

IDENTIFICATION AND MONITORING

1. Ecosystems and habitats: containing high diversity, large numbers of endemic or threatened species, or wilderness; required by migratory species; of social, economic, cultural or scientific importance; or, which are representative, unique or associated with key evolutionary or other biological processes;
2. Species and communities which are: threatened; wild relatives of domesticated or cultivated species; of medicinal, agricultural or other economic value; or social, scientific or cultural importance; or importance for research into the conservation and sustainable use of biological diversity, such as indicator species; and
3. Described genomes and genes of social, scientific or economic importance.

Annex II

Part 1

ARBITRATION

Article 1

The claimant party shall notify the secretariat that the parties are referring a dispute to arbitration pursuant to Article 27. The notification shall state the subject-matter of arbitration and include, in particular, the articles of the Convention or the protocol, the interpretation or application of which are at issue. If the parties do not agree on the subject matter of the dispute before the President of the tribunal is designated, the arbitral tribunal shall determine the subject matter. The secretariat shall forward the information thus received to all Contracting Parties to this Convention or to the protocol concerned.

Article 2

1. In disputes between two parties, the arbitral tribunal shall consist of three members. Each of the parties to the dispute shall appoint an arbitrator and the two arbitrators so appointed shall designate by common agreement the third arbitrator who shall be the President of the tribunal. The latter shall not be a national of one of the parties to the dispute, nor have his or her usual place of residence in the territory of one of these parties, nor be employed by any of them, nor have dealt with the case in any other capacity.

2. In disputes between more than two parties, parties in the same interest shall appoint one arbitrator jointly by agreement.

3. Any vacancy shall be filled in the manner prescribed for the initial appointment.

Article 3

1. If the President of the arbitral tribunal has not been designated within two months of the appointment of the second arbitrator, the Secretary-General of the United Nations shall, at the request of a party, designate the President within a further two-month period.

2. If one of the parties to the dispute does not appoint an arbitrator within two months of receipt of the request, the other party may inform the Secretary-General who shall make the designation within a further two-month period.

Article 4

The arbitral tribunal shall render its decisions in accordance with the provisions of this Convention, any protocols concerned, and international law.

Article 5

Unless the parties to the dispute otherwise agree, the arbitral tribunal shall determine its own rules of procedure.

Article 6

The arbitral tribunal may, at the request of one of the parties, recommend essential interim measures of protection.

Article 7

The parties to the dispute shall facilitate the work of the arbitral tribunal and, in particular, using all means at their disposal, shall:

(a) Provide it with all relevant documents, information and facilities; and

(b) Enable it, when necessary, to call witnesses or experts and receive their evidence.

Article 8

The parties and the arbitrators are under an obligation to protect the confidentiality of any information they receive in confidence during the proceedings of the arbitral tribunal.

Article 9

Unless the arbitral tribunal determines otherwise because of the particular circumstances of the case, the costs of the tribunal shall be borne by the parties to the dispute in equal shares. The tribunal shall keep a record of all its costs, and shall furnish a final statement thereof to the parties.

Article 10

Any Contracting Party that has an interest of a legal nature in the subject-matter of the dispute which may be affected by the decision in the case, may intervene in the proceedings with the consent of the tribunal.

Article 11

The tribunal may hear and determine counterclaims arising directly out of the subject-matter of the dispute.

Article 12

Decisions both on procedure and substance of the arbitral tribunal shall be taken by a majority vote of its members.

Article 13

If one of the parties to the dispute does not appear before the arbitral tribunal or fails to defend its case, the other party may request the tribunal to continue the proceedings and to make its award. Absence of a party or a failure of a party to defend its case shall not constitute a bar to the proceedings. Before rendering its final decision, the arbitral tribunal must satisfy itself that the claim is well founded in fact and law.

Article 14

The tribunal shall render its final decision within five months of the date on which it is fully constituted unless it finds it necessary to extend the time-limit for a period which should not exceed five more months.

Article 15

The final decision of the arbitral tribunal shall be confined to the subject-matter of the dispute and shall state the reasons on which it is based. It shall contain the names of the members who have participated and the date of the final decision. Any member of the tribunal may attach a separate or dissenting opinion to the final decision.

Article 16

The award shall be binding on the parties to the dispute. It shall be without appeal unless the parties to the dispute have agreed in advance to an appellate procedure.

Article 17

Any controversy which may arise between the parties to the dispute as regards the interpretation or manner of implementation of the final decision may be submitted by either party for decision to the arbitral tribunal which rendered it.

Part 2

CONCILIATION

Article 1

A conciliation commission shall be created upon the request of one of the parties to the dispute. The commission shall, unless the parties otherwise agree, be composed of five members, two appointed by each Party concerned and a President chosen jointly by those members.

Article 2

In disputes between more than two parties, parties in the same interest shall appoint their members of the commission jointly by agreement. Where two or more parties have separate interests or there is a disagreement as to whether they are of the same interest, they shall appoint their members separately.

Article 3

If any appointments by the parties are not made within two months of the date of the request to create a conciliation commission, the Secretary-General of the United Nations shall, if asked to do so by the party that made the request, make those appointments within a further two-month period.

Article 4

If a President of the conciliation commission has not been chosen within two months of the last of the members of the commission being appointed, the Secretary-General of the United Nations shall, if asked to do so by a party, designate a President within a further two-month period.

Article 5

The conciliation commission shall take its decisions by majority vote of its members. It shall, unless the parties to the dispute otherwise agree, determine its own procedure. It shall render a proposal for resolution of the dispute, which the parties shall consider in good faith.

Article 6

A disagreement as to whether the conciliation commission has competence shall be decided by the commission.

SIGNATORIES OF THE CONVENTION ON BIOLOGICAL DIVERSITY AT THE TIME OF THE UNITED NATIONS CONFERENCE ON ENVIRONMENT AND DEVELOPMENT

(RIO DE JANEIRO, 3–14 JUNE 1992)

	Signatory	*Date of signature*
1.	Antigua and Barbuda	5 June 1992
2.	Australia	5 June 1992
3.	Bangladesh	5 June 1992
4.	Belgium	5 June 1992
5.	Brazil	5 June 1992
6.	Finland	5 June 1992
7.	India	5 June 1992
8.	Indonesia	5 June 1992
9.	Italy	5 June 1992
10.	Liechtenstein	5 June 1992
11.	Republic of Moldova	5 June 1992
12.	Nauru	5 June 1992
13.	Netherlands	5 June 1992
14.	Pakistan	5 June 1992
15.	Poland	5 June 1992
16.	Romania	5 June 1992
17.	Bostwana	8 June 1992
18.	Madagascar	8 June 1992
19.	Sweden	8 June 1992
20.	Tuvalu	8 June 1992
21.	Yugoslavia	8 June 1992
22.	Bahrain	9 June 1992
23.	Ecuador	9 June 1992

24.	Egypt	9 June 1992
25.	Kazakhstan	9 June 1992
26.	Kuwait	9 June 1992
27.	Luxembourg	9 June 1992
28.	Norway	9 June 1992
29.	Sudan	9 June 1992
30.	Uruguay	9 June 1992
31.	Vanuatu	9 June 1992
32.	Cote d'Ivoire	10 June 1992
33.	Ethiopia	10 June 1992
34.	Iceland	10 June 1992
35.	Malawi	10 June 1992
36.	Mauritius	10 June 1992
37.	Oman	10 June 1992
38.	Rwanda	10 June 1992
39.	San Marino	10 June 1992
40.	Seychelles	10 June 1992
41.	Sri Lanka	10 June 1992
42.	Belarus	11 June 1992
43.	Bhutan	11 June 1992
44.	Burundi	11 June 1992
45.	Canada	11 June 1992
46.	China	11 June 1992
47.	Comoros	11 June 1992
48.	Congo	11 June 1992
49.	Croatia	11 June 1992
50.	Democratic People's Republic of Korea	11 June 1992
51.	Israel	11 June 1992
52.	Jamaica	11 June 1992
53.	Jordan	11 June 1992
54.	Kenya	11 June 1992
55.	Latvia	11 June 1992
56.	Lesotho	11 June 1992
57.	Lithuania	11 June 1992
58.	Monaco	11 June 1992
59.	Myanmar	11 June 1992
60.	Niger	11 June 1992
61.	Qatar	11 June 1992
62.	Trinidad and Tobago	11 June 1992
63.	Turkey	11 June 1992
64.	Ukraine	11 June 1992
65.	United Arab Emirates	11 June 1992
66.	Zaire	11 June 1992

67.	Zambia	11 June 1992
68.	Afghanistan	12 June 1992
69.	Angola	12 June 1992
70.	Argentina	12 June 1992
71.	Azerbaijan	12 June 1992
72.	Bahamas	12 June 1992
73.	Barbados	12 June 1992
74.	Bulgaria	12 June 1992
75.	Burkina Faso	12 June 1992
76.	Cape Verde	12 June 1992
77.	Chad	12 June 1992
78.	Colombia	12 June 1992
79.	Cook Islands	12 June 1992
80.	Cuba	12 June 1992
81.	Cyprus	12 June 1992
82.	Denmark	12 June 1992
83.	Estonia	12 June 1992
84.	Gabon	12 June 1992
85.	Gambia	12 June 1992
86.	Germany	12 June 1992
87.	Ghana	12 June 1992
88.	Greece	12 June 1992
89.	Guinea	12 June 1992
90.	Guinea-Bissau	12 June 1992
91.	Lebanon	12 June 1992
92.	Liberia	12 June 1992
93.	Malaysia	12 June 1992
94.	Maldives	12 June 1992
95.	Malta	12 June 1992
96.	Marshall Islands	12 June 1992
97.	Mauritania	12 June 1992
98.	Micronesia	12 June 1992
99.	Mongolia	12 June 1992
100.	Mozambique	12 June 1992
101.	Namibia	12 June 1992
102.	Nepal	12 June 1992
103.	New Zealand	12 June 1992
104.	Paraguay	12 June 1992
105.	Peru	12 June 1992
106.	Philippines	12 June 1992
107.	Saint Kitts and Nevis	12 June 1992
108.	Samoa	12 June 1992

109.	Sao Tome and Principe	12 June 1992
110.	Swaziland	12 June 1992
111.	Switzerland	12 June 1992
112.	Thailand	12 June 1992
113.	Togo	12 June 1992
114.	Uganda	12 June 1992
115.	United Kingdom of Great Britain and Northern Ireland	12 June 1992
116.	United Republic of Tanzania	12 June 1992
117.	Venezuela	12 June 1992
118.	Yemen	12 June 1992
119.	Zimbabwe	12 June 1992
120.	Algeria	13 June 1992
121.	Armenia	13 June 1992
122.	Austria	13 June 1992
123.	Belize	13 June 1992
124.	Benin	13 June 1992
125.	Bolivia	13 June 1992
126.	Central African Republic	13 June 1992
127.	Chile	13 June 1992
128.	Costa Rica	13 June 1992
129.	Djibouti	13 June 1992
130.	Dominican Republic	13 June 1992
131.	El Salvador	13 June 1992
132.	European Economic Community	13 June 1992
133.	France	13 June 1992
134.	Guatemala	13 June 1992
135.	Guyana	13 June 1992
136.	Haiti	13 June 1992
137.	Hungary	13 June 1992
138.	Honduras	13 June 1992
139.	Ireland	13 June 1992
140.	Japan	13 June 1992
141.	Mexico	13 June 1992
142.	Morocco	13 June 1992
143.	Nicaragua	13 June 1992
144.	Nigeria	13 June 1992
145.	Panama	13 June 1992
146.	Papua New Guinea	13 June 1992
147.	Portugal	13 June 1992
148.	Republic of Korea	13 June 1992
149.	Russian Federation	13 June 1992

150.	Senegal	13 June 1992
151.	Slovenia	13 June 1992
152.	Solomon Islands	13 June 1992
153.	Spain	13 June 1992
154.	Suriname	13 June 1992
155.	Tunisia	13 June 1992
156.	Cameroon	14 June 1992
157.	Iran	14 June 1992

APPENDIX III

ANNEX 1C

AGREEMENT ON TRADE-RELATED ASPECTS OF INTELLECTUAL PROPERTY RIGHTS

PART I GENERAL PROVISIONS AND BASIC PRINCIPLES

PART II STANDARDS CONCERNING THE AVAILABILITY, SCOPE AND USE OF INTELLECTUAL PROPERTY RIGHTS

1. Copyright and Related Rights
2. Trademarks
3. Geographical Indications
4. Industrial Designs
5. Patents
6. Layout-Designs (Topographies) of Integrated Circuits
7. Protection of Undisclosed Information
8. Control of Anti-Competitive Practices in Contractual Licences

AGREEMENT ON TRADE-RELATED ASPECTS OF INTELLECTUAL PROPERTY RIGHTS

Members,

Desiring to reduce distortions and impediments to international trade, and taking into account the need to promote effective and adequate protection of intellectual property rights, and to ensure that measures and procedures to enforce intellectual property rights do not themselves become barriers to legitimate trade;

Recognizing, to this end, the need for new rules and disciplines concerning:

(a) the applicability of the basic principles of GATT 1994 and of relevant international intellectual property agreements or conventions;
(b) the provision of adequate standards and principles concerning the availability, scope and use of trade-related intellectual property rights;
(c) the provision of effective and appropriate means for the enforcement of trade-related intellectual property rights, taking into account differences in national legal systems;
(d) the provision of effective and expeditious procedures for the multilateral prevention and settlement of disputes between governments; and
(e) transitional arrangements aiming at the fullest participation in the results of the negotiations;

Recognizing the need for a multilateral framework of principles, rules and disciplines dealing with international trade in counterfeit goods;

Recognizing that intellectual property rights are private rights;

Recognizing the underlying public policy objectives of national systems for the protection of intellectual property, including developmental and technological objectives;

Recognizing also the special needs of the least-developed country Members in respect of maximum flexibility in the domestic implementation of laws and regulations in order to enable them to create a sound and viable technological base;

Emphasizing the importance of reducing tensions by reaching strengthened commitments to resolve disputes on trade-related intellectual property issues through multilateral procedures;

Desiring to establish a mutually supportive relationship between the WTO and the World Intellectual Property Organization (referred to in this Agreement as "WIPO") as well as other relevant international organizations;

Hereby agree as follows:

PART I
GENERAL PROVISIONS AND BASIC PRINCIPLES

Article 1
Nature and Scope of Obligations

1. Members shall give effect to the provisions of this Agreement. Members may, but shall not be obliged to, implement in their law more extensive protection than is required by this Agreement, provided that such protection does not contravene the provisions of this Agreement. Members shall be free to determine the appropriate method of implementing the provisions of this Agreement within their own legal system and practice.

2. For the purposes of this Agreement, the term "intellectual property" refers to all categories of intellectual property that are the subject of Sections 1 through 7 of Part II.

3. Members shall accord the treatment provided for in this Agreement to the nationals of other Members.[1] In respect of the relevant intellectual property right, the nationals of other Members shall be understood as

1. When "nationals" are referred to in this Agreement, they shall be deemed, in the case of a separate customs territory Member of the WTO, to mean persons, natural or legal, who are domiciled or who have a real and effective industrial or commercial establishment in that customs territory.

those natural or legal persons that would meet the criteria for eligibility for protection provided for in the Paris Convention (1967), the Berne Convention (1971), the Rome Convention and the Treaty on Intellectual Property in Respect of Integrated Circuits, were all Members of the WTO members of those conventions.[2] Any Member availing itself of the possibilities provided in paragraph 3 of Article 5 or paragraph 2 of Article 6 of the Rome Convention shall make a notification as foreseen in those provisions to the Council for Trade-Related Aspects of Intellectual Property Rights (the "Council for TRIPS").

Article 2
Intellectual Property Conventions

1. In respect of Parts II, III and IV of this Agreement, Members shall comply with Articles 1 through 12, and Article 19, of the Paris Convention (1967).

2. Nothing in Parts I to IV of this Agreement shall derogate from existing obligations that Members may have to each other under the Paris Convention, the Berne Convention, the Rome Convention and the Treaty on Intellectual Property in Respect of Integrated Circuits.

Article 3
National Treatment

1. Each Member shall accord to the nationals of other Members treatment no less favourable than that it accords to its own nationals with regard to the protection[3] of intellectual property, subject to the exceptions already provided in, respectively, the Paris Convention (1967), the Berne Convention (1971), the Rome Convention or the Treaty on Intellectual Property in Respect of Integrated Circuits. In respect of performers,

2. In this Agreement, "Paris Convention" refers to the Paris Convention for the Protection of Industrial Property; "Paris Convention (1967)" refers to the Stockholm Act of this Convention of 14 July 1967. "Berne Convention" refers to the Berne Convention for the Protection of Literary and Artistic Works; "Berne Convention (1971)" refers to the Paris Act of this Convention of 24 July 1971. "Rome Convention" refers to the International Convention for the Protection of Performers, Producers of Phonograms and Broadcasting Organizations, adopted at Rome on 26 October 1961. "Treaty on Intellectual Property in Respect of Integrated Circuits" (IPIC Treaty) refers to the Treaty on Intellectual Property in Respect of Integrated Circuits, adopted at Washington on 26 May 1989. "WTO Agreement" refers to the Agreement Establishing the WTO.

3. For the purposes of Articles 3 and 4, "protection" shall include matters affecting the availability, acquisition, scope, maintenance and enforcement of intellectual property rights as well as those matters affecting the use of intellectual property rights specifically addressed in this Agreement.

producers of phonograms and broadcasting organizations, this obligation only applies in respect of the rights provided under this Agreement. Any Member availing itself of the possibilities provided in Article 6 of the Berne Convention (1971) or paragraph 1(b) of Article 16 of the Rome Convention shall make a notification as foreseen in those provisions to the Council for TRIPS.

2. Members may avail themselves of the exceptions permitted under paragraph 1 in relation to judicial and administrative procedures, including the designation of an address for service or the appointment of an agent within the jurisdiction of a Member, only where such exceptions are necessary to secure compliance with laws and regulations which are not inconsistent with the provisions of this Agreement and where such practices are not applied in a manner which would constitute a disguised restriction on trade.

Article 4
Most-Favoured-Nation Treatment

With regard to the protection of intellectual property, any advantage, favour, privilege or immunity granted by a Member to the nationals of any other country shall be accorded immediately and unconditionally to the nationals of all other Members. Exempted from this obligation are any advantage, favour, privilege or immunity accorded by a Member:

(a) deriving from international agreements on judicial assistance or law enforcement of a general nature and not particularly confined to the protection of intellectual property;

(b) granted in accordance with the provisions of the Berne Convention (1971) or the Rome Convention authorizing that the treatment accorded be a function not of national treatment but of the treatment accorded in another country;

(c) in respect of the rights of performers, producers of phonograms and broadcasting organizations not provided under this Agreement;

(d) deriving from international agreements related to the protection of intellectual property which entered into force prior to the entry into force of the WTO Agreement, provided that such agreements are notified to the Council for TRIPS and do not constitute an arbitrary or unjustifiable discrimination against nationals of other Members.

Article 5
Multilateral Agreements on Acquisition or Maintenance of Protection

The obligations under Articles 3 and 4 do not apply to procedures provided in multilateral agreements concluded under the auspices of WIPO relating to the acquisition or maintenance of intellectual property rights.

Article 6
Exhaustion

For the purposes of dispute settlement under this Agreement, subject to the provisions of Articles 3 and 4 nothing in this Agreement shall be used to address the issue of the exhaustion of intellectual property rights.

Article 7
Objectives

The protection and enforcement of intellectual property rights should contribute to the promotion of technological innovation and to the transfer and dissemination of technology, to the mutual advantage of producers and users of technological knowledge and in a manner conducive to social and economic welfare, and to a balance of rights and obligations.

Article 8
Principles

1. Members may, in formulating or amending their laws and regulations, adopt measures necessary to protect public health and nutrition, and to promote the public interest in sectors of vital importance to their socio-economic and technological development, provided that such measures are consistent with the provisions of this Agreement.

2. Appropriate measures, provided that they are consistent with the provisions of this Agreement, may be needed to prevent the abuse of intellectual property rights by right holders or the resort to practices which unreasonably restrain trade or adversely affect the international transfer of technology.

PART II
STANDARDS CONCERNING THE AVAILABILITY, SCOPE AND USE OF INTELLECTUAL PROPERTY RIGHTS

SECTION 1: COPYRIGHT AND RELATED RIGHTS

Article 9
Relation to the Berne Convention

1. Members shall comply with Articles 1 through 21 of the Berne Convention (1971) and the Appendix thereto. However, Members shall not have rights or obligations under this Agreement in respect of the rights conferred under Article 6*bis* of that Convention or of the rights derived therefrom.

2. Copyright protection shall extend to expressions and not to ideas, procedures, methods of operation or mathematical concepts as such.

Article 10
Computer Programs and Compilations of Data

1. Computer programs, whether in source or object code, shall be protected as literary works under the Berne Convention (1971).

2. Compilations of data or other material, whether in machine readable or other form, which by reason of the selection or arrangement of their contents constitute intellectual creations shall be protected as such. Such protection, which shall not extend to the data or material itself, shall be without prejudice to any copyright subsisting in the data or material itself.

Article 11
Rental Rights

In respect of at least computer programs and cinematographic works, a Member shall provide authors and their successors in title the right to authorize or to prohibit the commercial rental to the public of originals or copies of their copyright works. A Member shall be excepted from this obligation in respect of cinematographic works unless such rental has led to widespread copying of such works which is materially impairing the exclusive right of reproduction conferred in that Member on authors and their successors in title. In respect of computer programs, this obligation does not apply to rentals where the program itself is not the essential object of the rental.

Article 12
Term of Protection

Whenever the term of protection of a work, other than a photographic work or a work of applied art, is calculated on a basis other than the life of a natural person, such term shall be no less than 50 years from the end of the calendar year of authorized publication, or, failing such authorized publication within 50 years from the making of the work, 50 years from the end of the calendar year of making.

Article 13
Limitations and Exceptions

Members shall confine limitations or exceptions to exclusive rights to certain special cases which do not conflict with a normal exploitation of the work and do not unreasonably prejudice the legitimate interests of the right holder.

Article 14
Protection of Performers, Producers of Phonograms (Sound Recordings) and Broadcasting Organizations

1. In respect of a fixation of their performance on a phonogram, performers shall have the possibility of preventing the following acts when undertaken without their authorization: the fixation of their unfixed performance and the reproduction of such fixation. Performers shall also have the possibility of preventing the following acts when undertaken without their authorization: the broadcasting by wireless means and the communication to the public of their live performance.

2. Producers of phonograms shall enjoy the right to authorize or prohibit the direct or indirect reproduction of their phonograms.

3. Broadcasting organizations shall have the right to prohibit the following acts when undertaken without their authorization: the fixation, the reproduction of fixations, and the rebroadcasting by wireless means of broadcasts, as well as the communication to the public of television broadcasts of the same. Where Members do not grant such rights to broadcasting organizations, they shall provide owners of copyright in the subject matter of broadcasts with the possibility of preventing the above acts, subject to the provisions of the Berne Convention (1971).

4. The provisions of Article 11 in respect of computer programs shall apply *mutatis mutandis* to producers of phonograms and any other right holders in phonograms as determined in a Member's law. If on 15 April 1994 a Member has in force a system of equitable remuneration of right holders in respect of the rental of phonograms, it may maintain such system provided that the commercial rental of phonograms is not giving rise to the material impairment of the exclusive rights of reproduction of right holders.

5. The term of the protection available under this Agreement to performers and producers of phonograms shall last at least until the end of a period of 50 years computed from the end of the calendar year in which the fixation was made or the performance took place. The term of protection granted pursuant to paragraph 3 shall last for at least 20 years from the end of the calendar year in which the broadcast took place.

6. Any Member may, in relation to the rights conferred under paragraphs 1, 2 and 3, provide for conditions, limitations, exceptions and reservations to the extent permitted by the Rome Convention. However, the provisions of Article 18 of the Berne Convention (1971) shall also apply,

mutatis mutandis, to the rights of performers and producers of phonograms in phonograms.

SECTION 2: TRADEMARKS

Article 15
Protectable Subject Matter

1. Any sign, or any combination of signs, capable of distinguishing the goods or services of one undertaking from those of other undertakings, shall be capable of constituting a trademark. Such signs, in particular words including personal names, letters, numerals, figurative elements and combinations of colours as well as any combination of such signs, shall be eligible for registration as trademarks. Where signs are not inherently capable of distinguishing the relevant goods or services, Members may make registrability depend on distinctiveness acquired through use. Members may require, as a condition of registration, that signs be visually perceptible.

2. Paragraph 1 shall not be understood to prevent a Member from denying registration of a trademark on other grounds, provided that they do not derogate from the provisions of the Paris Convention (1967).

3. Members may make registrability depend on use. However, actual use of a trademark shall not be a condition for filing an application for registration. An application shall not be refused solely on the ground that intended use has not taken place before the expiry of a period of three years from the date of application.

4. The nature of the goods or services to which a trademark is to be applied shall in no case form an obstacle to registration of the trademark.

5. Members shall publish each trademark either before it is registered or promptly after it is registered and shall afford a reasonable opportunity for petitions to cancel the registration. In addition, Members may afford an opportunity for the registration of a trademark to be opposed.

Article 16
Rights Conferred

1. The owner of a registered trademark shall have the exclusive right to prevent all third parties not having the owner's consent from using in the course of trade identical or similar signs for goods or services which are identical or similar to those in respect of which the trademark is registered where such use would result in a likelihood of confusion. In case of the use of an identical sign for identical goods or services, a likelihood of confusion shall be presumed. The rights described above shall not

prejudice any existing prior rights, nor shall they affect the possibility of Members making rights available on the basis of use.

2. Article 6*bis* of the Paris Convention (1967) shall apply, *mutatis mutandis*, to services. In determining whether a trademark is well-known, Members shall take account of the knowledge of the trademark in the relevant sector of the public, including knowledge in the Member concerned which has been obtained as a result of the promotion of the trademark.

3. Article 6*bis* of the Paris Convention (1967) shall apply, *mutatis mutandis*, to goods or services which are not similar to those in respect of which a trademark is registered, provided that use of that trademark in relation to those goods or services would indicate a connection between those goods or services and the owner of the registered trademark and provided that the interests of the owner of the registered trademark are likely to be damaged by such use.

Article 17
Exceptions

Members may provide limited exceptions to the rights conferred by a trademark, such as fair use of descriptive terms, provided that such exceptions take account of the legitimate interests of the owner of the trademark and of third parties.

Article 18
Term of Protection

Initial registration, and each renewal of registration, of a trademark shall be for a term of no less than seven years. The registration of a trademark shall be renewable indefinitely.

Article 19
Requirement of Use

1. If use is required to maintain a registration, the registration may be cancelled only after an uninterrupted period of at least three years of non-use, unless valid reasons based on the existence of obstacles to such use are shown by the trademark owner. Circumstances arising independently of the will of the owner of the trademark which constitute an obstacle to the use of the trademark, such as import restrictions on or other government requirements for goods or services protected by the trademark, shall be recognized as valid reasons for non-use.

2. When subject to the control of its owner, use of a trademark by an-

other person shall be recognized as use of the trademark for the purpose of maintaining the registration.

Article 20
Other Requirements

The use of a trademark in the course of trade shall not be unjustifiably encumbered by special requirements, such as use with another trademark, use in a special form or use in a manner detrimental to its capability to distinguish the goods or services of one undertaking from those of other undertakings. This will not preclude a requirement prescribing the use of the trademark identifying the undertaking producing the goods or services along with, but without linking it to, the trademark distinguishing the specific goods or services in question of that undertaking.

Article 21
Licensing and Assignment

Members may determine conditions on the licensing and assignment of trademarks, it being understood that the compulsory licensing of trademarks shall not be permitted and that the owner of a registered trademark shall have the right to assign the trademark with or without the transfer of the business to which the trademark belongs.

SECTION 3: GEOGRAPHICAL INDICATIONS

Article 22
Protection of Geographical Indications

1. Geographical indications are, for the purposes of this Agreement, indications which identify a good as originating in the territory of a Member, or a region or locality in that territory, where a given quality, reputation or other characteristic of the good is essentially attributable to its geographical origin.

2. In respect of geographical indications, Members shall provide the legal means for interested parties to prevent:
(a) the use of any means in the designation or presentation of a good that indicates or suggests that the good in question originates in a geographical area other than the true place of origin in a manner which misleads the public as to the geographical origin of the good;
(b) any use which constitutes an act of unfair competition within the meaning of Article 10*bis* of the Paris Convention (1967).

3. A Member shall, *ex officio* if its legislation so permits or at the request of an interested party, refuse or invalidate the registration of a

trademark which contains or consists of a geographical indication with respect to goods not originating in the territory indicated, if use of the indication in the trademark for such goods in that Member is of such a nature as to mislead the public as to the true place of origin.

4. The protection under paragraphs 1, 2 and 3 shall be applicable against a geographical indication which, although literally true as to the territory, region or locality in which the goods originate, falsely represents to the public that the goods originate in another territory.

Article 23
Additional Protection for Geographical Indications for Wines and Spirits

1. Each Member shall provide the legal means for interested parties to prevent use of a geographical indication identifying wines for wines not originating in the place indicated by the geographical indication in question or identifying spirits for spirits not originating in the place indicated by the geographical indication in question, even where the true origin of the goods is indicated or the geographical indication is used in translation or accompanied by expressions such as "kind", "type", "style", "imitation" or the like.[4]

2. The registration of a trademark for wines which contains or consists of a geographical indication identifying wines or for spirits which contains or consists of a geographical indication identifying spirits shall be refused or invalidated, *ex officio* if a Member's legislation so permits or at the request of an interested party, with respect to such wines or spirits not having this origin.

3. In the case of homonymous geographical indications for wines, protection shall be accorded to each indication, subject to the provisions of paragraph 4 of Article 22. Each Member shall determine the practical conditions under which the homonymous indications in question will be differentiated from each other, taking into account the need to ensure equitable treatment of the producers concerned and that consumers are not misled.

4. In order to facilitate the protection of geographical indications for wines, negotiations shall be undertaken in the Council for TRIPS concerning the establishment of a multilateral system of notification and registration of geographical indications for wines eligible for protection in those Members participating in the system.

4. Notwithstanding the first sentence of Article 42, Members may, with respect to these obligations, instead provide for enforcement by administrative action.

Article 24
International Negotiations; Exceptions

1. Members agree to enter into negotiations aimed at increasing the protection of individual geographical indications under Article 23. The provisions of paragraphs 4 through 8 below shall not be used by a Member to refuse to conduct negotiations or to conclude bilateral or multilateral agreements. In the context of such negotiations, Members shall be willing to consider the continued applicability of these provisions to individual geographical indications whose use was the subject of such negotiations.

2. The Council for TRIPS shall keep under review the application of the provisions of this Section; the first such review shall take place within two years of the entry into force of the WTO Agreement. Any matter affecting the compliance with the obligations under these provisions may be drawn to the attention of the Council, which, at the request of a Member, shall consult with any Member or Members in respect of such matter in respect of which it has not been possible to find a satisfactory solution through bilateral or plurilateral consultations between the Members concerned. The Council shall take such action as may be agreed to facilitate the operation and further the objectives of this Section.

3. In implementing this Section, a Member shall not diminish the protection of geographical indications that existed in that Member immediately prior to the date of entry into force of the WTO Agreement.

4. Nothing in this Section shall require a Member to prevent continued and similar use of a particular geographical indication of another Member identifying wines or spirits in connection with goods or services by any of its nationals or domiciliaries who have used that geographical indication in a continuous manner with regard to the same or related goods or services in the territory of that Member either (*a*) for at least 10 years preceding 15 April 1994 or (*b*) in good faith preceding that date.

5. Where a trademark has been applied for or registered in good faith, or where rights to a trademark have been acquired through use in good faith either:

(a) before the date of application of these provisions in that Member as defined in Part VI; or
(b) before the geographical indication is protected in its country of origin;

measures adopted to implement this Section shall not prejudice eligibility for or the validity of the registration of a trademark, or the right to use a trademark, on the basis that such a trademark is identical with, or similar to, a geographical indication.

6. Nothing in this Section shall require a Member to apply its provisions in respect of a geographical indication of any other Member with respect to goods or services for which the relevant indication is identical with the term customary in common language as the common name for such goods or services in the territory of that Member. Nothing in this Section shall require a Member to apply its provisions in respect of a geographical indication of any other Member with respect to products of the vine for which the relevant indication is identical with the customary name of a grape variety existing in the territory of that Member as of the date of entry into force of the WTO Agreement.

7. A Member may provide that any request made under this Section in connection with the use or registration of a trademark must be presented within five years after the adverse use of the protected indication has become generally known in that Member or after the date of registration of the trademark in that Member provided that the trademark has been published by that date, if such date is earlier than the date on which the adverse use became generally known in that Member, provided that the geographical indication is not used or registered in bad faith.

8. The provisions of this Section shall in no way prejudice the right of any person to use, in the course of trade, that person's name or the name of that person's predecessor in business, except where such name is used in such a manner as to mislead the public.

9. There shall be no obligation under this Agreement to protect geographical indications which are not or cease to be protected in their country of origin, or which have fallen into disuse in that country.

SECTION 4: INDUSTRIAL DESIGNS

Article 25
Requirements for Protection

1. Members shall provide for the protection of independently created industrial designs that are new or original. Members may provide that designs are not new or original if they do not significantly differ from known designs or combinations of known design features. Members may provide that such protection shall not extend to designs dictated essentially by technical or functional considerations.

2. Each Member shall ensure that requirements for securing protection for textile designs, in particular in regard to any cost, examination or publication, do not unreasonably impair the opportunity to seek and obtain such protection. Members shall be free to meet this obligation through industrial design law or through copyright law.

Article 26
Protection

1. The owner of a protected industrial design shall have the right to prevent third parties not having the owner's consent from making, selling or importing articles bearing or embodying a design which is a copy, or substantially a copy, of the protected design, when such acts are undertaken for commercial purposes.

2. Members may provide limited exceptions to the protection of industrial designs, provided that such exceptions do not unreasonably conflict with the normal exploitation of protected industrial designs and do not unreasonably prejudice the legitimate interests of the owner of the protected design, taking account of the legitimate interests of third parties.

3. The duration of protection available shall amount to at least 10 years.

SECTION 5: PATENTS

Article 27
Patentable Subject Matter

1. Subject to the provisions of paragraphs 2 and 3, patents shall be available for any inventions, whether products or processes, in all fields of technology, provided that they are new, involve an inventive step and are capable of industrial application.[5] Subject to paragraph 4 of Article 65, paragraph 8 of Article 70 and paragraph 3 of this Article, patents shall be available and patent rights enjoyable without discrimination as to the place of invention, the field of technology and whether products are imported or locally produced.

2. Members may exclude from patentability inventions, the prevention within their territory of the commercial exploitation of which is necessary to protect *ordre public* or morality, including to protect human, animal or plant life or health or to avoid serious prejudice to the environment, provided that such exclusion is not made merely because the exploitation is prohibited by their law.

3. Members may also exclude from patentability:

(a) diagnostic, therapeutic and surgical methods for the treatment of humans or animals;

(b) plants and animals other than micro-organisms, and essentially biological processes for the production of plants or animals other than

5. For the purposes of this Article, the terms "inventive step" and "capable of industrial application" may be deemed by a Member to be synonymous with the terms "non-obvious" and "useful" respectively.

non-biological and microbiological processes. However, Members shall provide for the protection of plant varieties either by patents or by an effective *sui generis* system or by any combination thereof. The provisions of this subparagraph shall be reviewed four years after the date of entry into force of the WTO Agreement.

Article 28
Rights Conferred

1. A patent shall confer on its owner the following exclusive rights:
(a) where the subject matter of a patent is a product, to prevent third parties not having the owner's consent from the acts of: making, using, offering for sale, selling, or importing[6] for these purposes that product;
(b) where the subject matter of a patent is a process, to prevent third parties not having the owner's consent from the act of using the process, and from the acts of: using, offering for sale, selling, or importing for these purposes at least the product obtained directly by that process.
2. Patent owners shall also have the right to assign, or transfer by succession, the patent and to conclude licensing contracts.

Article 29
Conditions on Patent Applicants

1. Members shall require that an applicant for a patent shall disclose the invention in a manner sufficiently clear and complete for the invention to be carried out by a person skilled in the art and may require the applicant to indicate the best mode for carrying out the invention known to the inventor at the filing date or, where priority is claimed, at the priority date of the application.
2. Members may require an applicant for a patent to provide information concerning the applicant's corresponding foreign applications and grants.

Article 30
Exceptions to Rights Conferred

Members may provide limited exceptions to the exclusive rights conferred by a patent, provided that such exceptions do not unreasonably conflict with a normal exploitation of the patent and do not unreasonably

6. This right, like all other rights conferred under this Agreement in respect of the use, sale, importation or other distribution of goods, is subject to the provisions of Article 6.

prejudice the legitimate interests of the patent owner, taking account of the legitimate interests of third parties.

Article 31
Other Use Without Authorization of the Right Holder

Where the law of a Member allows for other use[7] of the subject matter of a patent without the authorization of the right holder, including use by the government or third parties authorized by the government, the following provisions shall be respected:

(a) authorization of such use shall be considered on its individual merits;
(b) such use may only be permitted if, prior to such use, the proposed user has made efforts to obtain authorization from the right holder on reasonable commercial terms and conditions and that such efforts have not been successful within a reasonable period of time. This requirement may be waived by a Member in the case of a national emergency or other circumstances of extreme urgency or in cases of public non-commercial use. In situations of national emergency or other circumstances of extreme urgency, the right holder shall, nevertheless, be notified as soon as reasonably practicable. In the case of public non-commercial use, where the government or contractor, without making a patent search, knows or has demonstrable grounds to know that a valid patent is or will be used by or for the government, the right holder shall be informed promptly;
(c) the scope and duration of such use shall be limited to the purpose for which it was authorized, and in the case of semi-conductor technology shall only be for public non-commercial use or to remedy a practice determined after judicial or administrative process to be anti-competitive;
(d) such use shall be non-exclusive;
(e) such use shall be non-assignable, except with that part of the enterprise or goodwill which enjoys such use;
(f) any such use shall be authorized predominantly for the supply of the domestic market of the Member authorizing such use;
(g) authorization for such use shall be liable, subject to adequate protection of the legitimate interests of the persons so authorized, to be terminated if and when the circumstances which led to it cease to exist and are unlikely to recur. The competent authority shall have the authority to review, upon motivated request, the continued existence of these circumstances;
(h) the right holder shall be paid adequate remuneration in the circum-

7. "Other use" refers to use other than that allowed under Article 30.

stances of each case, taking into account the economic value of the authorization;

(i) the legal validity of any decision relating to the authorization of such use shall be subject to judicial review or other independent review by a distinct higher authority in that Member;

(j) any decision relating to the remuneration provided in respect of such use shall be subject to judicial review or other independent review by a distinct higher authority in that Member;

(k) Members are not obliged to apply the conditions set forth in subparagraphs (b) and (f) where such use is permitted to remedy a practice determined after judicial or administrative process to be anti-competitive. The need to correct anti-competitive practices may be taken into account in determining the amount of remuneration in such cases. Competent authorities shall have the authority to refuse termination of authorization if and when the conditions which led to such authorization are likely to recur;

(l) where such use is authorized to permit the exploitation of a patent ("the second patent") which cannot be exploited without infringing another patent ("the first patent"), the following additional conditions shall apply:

    (i) the invention claimed in the second patent shall involve an important technical advance of considerable economic significance in relation to the invention claimed in the first patent;

    (ii) the owner of the first patent shall be entitled to a cross-licence on reasonable terms to use the invention claimed in the second patent; and

    (iii) the use authorized in respect of the first patent shall be non-assignable except with the assignment of the second patent.

Article 32
Revocation/Forfeiture

An opportunity for judicial review of any decision to revoke or forfeit a patent shall be available.

Article 33
Term of Protection

The term of protection available shall not end before the expiration of a period of twenty years counted from the filing date.[8]

8. It is understood that those Members which do not have a system of original grant may provide that the term of protection shall be computed from the filing date in the system of original grant.

Article 34
Process Patents: Burden of Proof

1. For the purposes of civil proceedings in respect of the infringement of the rights of the owner referred to in paragraph 1(b) of Article 28, if the subject matter of a patent is a process for obtaining a product, the judicial authorities shall have the authority to order the defendant to prove that the process to obtain an identical product is different from the patented process. Therefore, Members shall provide, in at least one of the following circumstances, that any identical product when produced without the consent of the patent owner shall, in the absence of proof to the contrary, be deemed to have been obtained by the patented process:
(a) if the product obtained by the patented process is new;
(b) if there is a substantial likelihood that the identical product was made by the process and the owner of the patent has been unable through reasonable efforts to determine the process actually used.

2. Any Member shall be free to provide that the burden of proof indicated in paragraph 1 shall be on the alleged infringer only if the condition referred to in subparagraph (a) is fulfilled or only if the condition referred to in subparagraph (b) is fulfilled.

3. In the adduction of proof to the contrary, the legitimate interests of defendants in protecting their manufacturing and business secrets shall be taken into account.

SECTION 6: LAYOUT-DESIGNS (TOPOGRAPHIES) OF INTEGRATED CIRCUITS

Article 35
Relation to the IPIC Treaty

Members agree to provide protection to the layout-designs (topographies) of integrated circuits (referred to in this Agreement as "layout-designs") in accordance with Articles 2 through 7 (other than paragraph 3 of Article 6), Article 12 and paragraph 3 of Article 16 of the Treaty on Intellectual Property in Respect of Integrated Circuits and, in addition, to comply with the following provisions.

Article 36
Scope of the Protection

Subject to the provisions of paragraph 1 of Article 37, Members shall consider unlawful the following acts if performed without the authorization of the right holder:[9] importing, selling, or otherwise distributing for commercial purposes a protected layout-design, an integrated circuit in which a protected layout-design is incorporated, or an article incorporating such an integrated circuit only in so far as it continues to contain an unlawfully reproduced layout-design.

Article 37
Acts Not Requiring the Authorization of the Right Holder

1. Notwithstanding Article 36, no Member shall consider unlawful the performance of any of the acts referred to in that Article in respect of an integrated circuit incorporating an unlawfully reproduced layout-design or any article incorporating such an integrated circuit where the person performing or ordering such acts did not know and had no reasonable ground to know, when acquiring the integrated circuit or article incorporating such an integrated circuit, that it incorporated an unlawfully reproduced layout-design. Members shall provide that, after the time that such person has received sufficient notice that the layout-design was unlawfully reproduced, that person may perform any of the acts with respect to the stock on hand or ordered before such time, but shall be liable to pay to the right holder a sum equivalent to a reasonable royalty such as would be payable under a freely negotiated licence in respect of such a layout-design.
2. The conditions set out in subparagraphs (a) through (k) of Article 31 shall apply *mutatis mutandis* in the event of any non-voluntary licensing of a layout-design or of its use by or for the government without the authorization of the right holder.

Article 38
Term of Protection

1. In Members requiring registration as a condition of protection, the term of protection of layout-designs shall not end before the expiration of a period of 10 years counted from the date of filing an application for registration or from the first commercial exploitation wherever in the world it occurs.
2. In Members not requiring registration as a condition for protection, layout-designs shall be protected for a term of no less than 10 years from the date of the first commercial exploitation wherever in the world it occurs.

9. The term "right holder" in this Section shall be understood as having the same meaning as the term "holder of the right" in the IPIC Treaty.

3. Notwithstanding paragraphs 1 and 2, a Member may provide that protection shall lapse 15 years after the creation of the layout-design.

SECTION 7: PROTECTION OF UNDISCLOSED INFORMATION

Article 39

1. In the course of ensuring effective protection against unfair competition as provided in Article 10*bis* of the Paris Convention (1967), Members shall protect undisclosed information in accordance with paragraph 2 and data submitted to governments or governmental agencies in accordance with paragraph 3.

2. Natural and legal persons shall have the possibility of preventing information lawfully within their control from being disclosed to, acquired by, or used by others without their consent in a manner contrary to honest commercial practices[10] so long as such information:

(a) is secret in the sense that it is not, as a body or in the precise configuration and assembly of its components, generally known among or readily accessible to persons within the circles that normally deal with the kind of information in question;
(b) has commercial value because it is secret; and
(c) has been subject to reasonable steps under the circumstances, by the person lawfully in control of the information, to keep it secret.

3. Members, when requiring, as a condition of approving the marketing of pharmaceutical or of agricultural chemical products which utilize new chemical entities, the submission of undisclosed test or other data, the origination of which involves a considerable effort, shall protect such data against unfair commercial use. In addition, Members shall protect such data against disclosure, except where necessary to protect the public, or unless steps are taken to ensure that the data are protected against unfair commercial use.

SECTION 8: CONTROL OF ANTI-COMPETITIVE PRACTICES IN CONTRACTUAL LICENCES

Article 40

1. Members agree that some licensing practices or conditions pertaining to intellectual property rights which restrain competition may have ad-

10. For the purpose of this provision, "a manner contrary to honest commercial practices" shall mean at least practices such as breach of contract, breach of confidence and inducement to breach, and includes the acquisition of undisclosed information by third parties who knew, or were grossly negligent in failing to know, that such practices were involved in the acquisition.

verse effects on trade and may impede the transfer and dissemination of technology.

2. Nothing in this Agreement shall prevent Members from specifying in their legislation licensing practices or conditions that may in particular cases constitute an abuse of intellectual property rights having an adverse effect on competition in the relevant market. As provided above, a Member may adopt, consistently with the other provisions of this Agreement, appropriate measures to prevent or control such practices, which may include for example exclusive grantback conditions, conditions preventing challenges to validity and coercive package licensing, in the light of the relevant laws and regulations of that Member.

3. Each Member shall enter, upon request, into consultations with any other Member which has cause to believe that an intellectual property right owner that is a national or domiciliary of the Member to which the request for consultations has been addressed is undertaking practices in violation of the requesting Member's laws and regulations on the subject matter of this Section, and which wishes to secure compliance with such legislation, without prejudice to any action under the law and to the full freedom of an ultimate decision of either Member. The Member addressed shall accord full and sympathetic consideration to, and shall afford adequate opportunity for, consultations with the requesting Member, and shall cooperate through supply of publicly available non-confidential information of relevance to the matter in question and of other information available to the Member, subject to domestic law and to the conclusion of mutually satisfactory agreements concerning the safeguarding of its confidentiality by the requesting Member.

4. A Member whose nationals or domiciliaries are subject to proceedings in another Member concerning alleged violation of that other Member's laws and regulations on the subject matter of this Section shall, upon request, be granted an opportunity for consultations by the other Member under the same conditions as those foreseen in paragraph 3.

PART III
ENFORCEMENT OF INTELLECTUAL PROPERTY RIGHTS

SECTION 1: GENERAL OBLIGATIONS

Article 41

1. Members shall ensure that enforcement procedures as specified in this Part are available under their law so as to permit effective action against any act of infringement of intellectual property rights covered by

this Agreement, including expeditious remedies to prevent infringements and remedies which constitute a deterrent to further infringements. These procedures shall be applied in such a manner as to avoid the creation of barriers to legitimate trade and to provide for safeguards against their abuse.

2. Procedures concerning the enforcement of intellectual property rights shall be fair and equitable. They shall not be unnecessarily complicated or costly, or entail unreasonable time-limits or unwarranted delays.

3. Decisions on the merits of a case shall preferably be in writing and reasoned. They shall be made available at least to the parties to the proceeding without undue delay. Decisions on the merits of a case shall be based only on evidence in respect of which parties were offered the opportunity to be heard.

4. Parties to a proceeding shall have an opportunity for review by a judicial authority of final administrative decisions and, subject to jurisdictional provisions in a Member's law concerning the importance of a case, of at least the legal aspects of initial judicial decisions on the merits of a case. However, there shall be no obligation to provide an opportunity for review of acquittals in criminal cases.

5. It is understood that this Part does not create any obligation to put in place a judicial system for the enforcement of intellectual property rights distinct from that for the enforcement of law in general, nor does it affect the capacity of Members to enforce their law in general. Nothing in this Part creates any obligation with respect to the distribution of resources as between enforcement of intellectual property rights and the enforcement of law in general.

SECTION 2: CIVIL AND ADMINISTRATIVE PROCEDURES AND REMEDIES

Article 42
Fair and Equitable Procedures

Members shall make available to right holders[11] civil judicial procedures concerning the enforcement of any intellectual property right covered by this Agreement. Defendants shall have the right to written notice which is timely and contains sufficient detail, including the basis of the claims. Parties shall be allowed to be represented by independent legal counsel, and procedures shall not impose overly burdensome requirements concerning mandatory personal appearances. All parties to such

11. For the purpose of this Part, the term "right holder" includes federations and associations having legal standing to assert such rights.

procedures shall be duly entitled to substantiate their claims and to present all relevant evidence. The procedure shall provide a means to identify and protect confidential information, unless this would be contrary to existing constitutional requirements.

Article 43
Evidence

1. The judicial authorities shall have the authority, where a party has presented reasonably available evidence sufficient to support its claims and has specified evidence relevant to substantiation of its claims which lies in the control of the opposing party, to order that this evidence be produced by the opposing party, subject in appropriate cases to conditions which ensure the protection of confidential information.

2. In cases in which a party to a proceeding voluntarily and without good reason refuses access to, or otherwise does not provide necessary information within a reasonable period, or significantly impedes a procedure relating to an enforcement action, a Member may accord judicial authorities the authority to make preliminary and final determinations, affirmative or negative, on the basis of the information presented to them, including the complaint or the allegation presented by the party adversely affected by the denial of access to information, subject to providing the parties an opportunity to be heard on the allegations or evidence.

Article 44
Injunctions

1. The judicial authorities shall have the authority to order a party to desist from an infringement, *inter alia* to prevent the entry into the channels of commerce in their jurisdiction of imported goods that involve the infringement of an intellectual property right, immediately after customs clearance of such goods. Members are not obliged to accord such authority in respect of protected subject matter acquired or ordered by a person prior to knowing or having reasonable grounds to know that dealing in such subject matter would entail the infringement of an intellectual property right.

2. Notwithstanding the other provisions of this Part and provided that the provisions of Part II specifically addressing use by governments, or by third parties authorized by a government, without the authorization of the right holder are complied with, Members may limit the remedies available against such use to payment of remuneration in accordance with subparagraph (h) of Article 31. In other cases, the remedies under this Part shall apply or, where these remedies are inconsistent with a

Member's law, declaratory judgments and adequate compensation shall be available.

Article 45
Damages

1. The judicial authorities shall have the authority to order the infringer to pay the right holder damages adequate to compensate for the injury the right holder has suffered because of an infringement of that person's intellectual property right by an infringer who knowingly, or with reasonable grounds to know, engaged in infringing activity.

2. The judicial authorities shall also have the authority to order the infringer to pay the right holder expenses, which may include appropriate attorney's fees. In appropriate cases, Members may authorize the judicial authorities to order recovery of profits and/or payment of pre-established damages even where the infringer did not knowingly, or with reasonable grounds to know, engage in infringing activity.

Article 46
Other Remedies

In order to create an effective deterrent to infringement, the judicial authorities shall have the authority to order that goods that they have found to be infringing be, without compensation of any sort, disposed of outside the channels of commerce in such a manner as to avoid any harm caused to the right holder, or, unless this would be contrary to existing constitutional requirements, destroyed. The judicial authorities shall also have the authority to order that materials and implements the predominant use of which has been in the creation of the infringing goods be, without compensation of any sort, disposed of outside the channels of commerce in such a manner as to minimize the risks of further infringements. In considering such requests, the need for proportionality between the seriousness of the infringement and the remedies ordered as well as the interests of third parties shall be taken into account. In regard to counterfeit trademark goods, the simple removal of the trademark unlawfully affixed shall not be sufficient, other than in exceptional cases, to permit release of the goods into the channels of commerce.

Article 47
Right of Information

Members may provide that the judicial authorities shall have the authority, unless this would be out of proportion to the seriousness of the infringement, to order the infringer to inform the right holder of the identity of third persons involved in the production and distribution of the infringing goods or services and of their channels of distribution.

Article 48
Indemnification of the Defendant

1. The judicial authorities shall have the authority to order a party at whose request measures were taken and who has abused enforcement procedures to provide to a party wrongfully enjoined or restrained adequate compensation for the injury suffered because of such abuse. The judicial authorities shall also have the authority to order the applicant to pay the defendant expenses, which may include appropriate attorney's fees.

2. In respect of the administration of any law pertaining to the protection or enforcement of intellectual property rights, Members shall only exempt both public authorities and officials from liability to appropriate remedial measures where actions are taken or intended in good faith in the course of the administration of that law.

Article 49
Administrative Procedures

To the extent that any civil remedy can be ordered as a result of administrative procedures on the merits of a case, such procedures shall conform to principles equivalent in substance to those set forth in this Section.

SECTION 3: PROVISIONAL MEASURES

Article 50

1. The judicial authorities shall have the authority to order prompt and effective provisional measures:

(a) to prevent an infringement of any intellectual property right from occurring, and in particular to prevent the entry into the channels of commerce in their jurisdiction of goods, including imported goods immediately after customs clearance;

(b) to preserve relevant evidence in regard to the alleged infringement.

2. The judicial authorities shall have the authority to adopt provisional measures *inaudita altera parte* where appropriate, in particular where any delay is likely to cause irreparable harm to the right holder, or where there is a demonstrable risk of evidence being destroyed.

3. The judicial authorities shall have the authority to require the applicant to provide any reasonably available evidence in order to satisfy themselves with a sufficient degree of certainty that the applicant is the right holder and that the applicant's right is being infringed or that such infringement is imminent, and to order the applicant to provide a security

or equivalent assurance sufficient to protect the defendant and to prevent abuse.

4. Where provisional measures have been adopted *inaudita altera parte*, the parties affected shall be given notice, without delay after the execution of the measures at the latest. A review, including a right to be heard, shall take place upon request of the defendant with a view to deciding, within a reasonable period after the notification of the measures, whether these measures shall be modified, revoked or confirmed.

5. The applicant may be required to supply other information necessary for the identification of the goods concerned by the authority that will execute the provisional measures.

6. Without prejudice to paragraph 4, provisional measures taken on the basis of paragraphs 1 and 2 shall, upon request by the defendant, be revoked or otherwise cease to have effect, if proceedings leading to a decision on the merits of the case are not initiated within a reasonable period, to be determined by the judicial authority ordering the measures where a Member's law so permits or, in the absence of such a determination, not to exceed 20 working days or 31 calendar days, whichever is the longer.

7. Where the provisional measures are revoked or where they lapse due to any act or omission by the applicant, or where it is subsequently found that there has been no infringement or threat of infringement of an intellectual property right, the judicial authorities shall have the authority to order the applicant, upon request of the defendant, to provide the defendant appropriate compensation for any injury caused by these measures.

8. To the extent that any provisional measure can be ordered as a result of administrative procedures, such procedures shall conform to principles equivalent in substance to those set forth in this Section.

SECTION 4: SPECIAL REQUIREMENTS RELATED TO BORDER MEASURES[12]

Article 51
Suspension of Release by Customs Authorities

Members shall, in conformity with the provisions set out below, adopt procedures[13] to enable a right holder, who has valid grounds for suspect-

12. Where a Member has dismantled substantially all controls over movement of goods across its border with another Member with which it forms part of a customs union, it shall not be required to apply the provisions of this Section at that border.

13. It is understood that there shall be no obligation to apply such procedures to imports of goods put on the market in another country by or with the consent of the right holder, or to goods in transit.

ing that the importation of counterfeit trademark or pirated copyright goods[14] may take place, to lodge an application in writing with competent authorities, administrative or judicial, for the suspension by the customs authorities of the release into free circulation of such goods. Members may enable such an application to be made in respect of goods which involve other infringements of intellectual property rights, provided that the requirements of this Section are met. Members may also provide for corresponding procedures concerning the suspension by the customs authorities of the release of infringing goods destined for exportation from their territories.

Article 52
Application

Any right holder initiating the procedures under Article 51 shall be required to provide adequate evidence to satisfy the competent authorities that, under the laws of the country of importation, there is *prima facie* an infringement of the right holder's intellectual property right and to supply a sufficiently detailed description of the goods to make them readily recognizable by the customs authorities. The competent authorities shall inform the applicant within a reasonable period whether they have accepted the application and, where determined by the competent authorities, the period for which the customs authorities will take action.

Article 53
Security or Equivalent Assurance

1. The competent authorities shall have the authority to require an applicant to provide a security or equivalent assurance sufficient to protect the defendant and the competent authorities and to prevent abuse. Such security or equivalent assurance shall not unreasonably deter recourse to these procedures.

2. Where pursuant to an application under this Section the release of goods involving industrial designs, patents, layout-designs or undisclosed

14. For the purposes of this Agreement:
(a) "counterfeit trademark goods" shall mean any goods, including packaging, bearing without authorization a trademark which is identical to the trademark validly registered in respect of such goods, or which cannot be distinguished in its essential aspects from such a trademark, and which thereby infringes the rights of the owner of the trademark in question under the law of the country of importation;
(b) "pirated copyright goods" shall mean any goods which are copies made without the consent of the right holder or person duly authorized by the right holder in the country of production and which are made directly or indirectly from an article where the making of that copy would have constituted an infringement of a copyright or a related right under the law of the country of importation.

information into free circulation has been suspended by customs authorities on the basis of a decision other than by a judicial or other independent authority, and the period provided for in Article 55 has expired without the granting of provisional relief by the duly empowered authority, and provided that all other conditions for importation have been complied with, the owner, importer, or consignee of such goods shall be entitled to their release on the posting of a security in an amount sufficient to protect the right holder for any infringement. Payment of such security shall not prejudice any other remedy available to the right holder, it being understood that the security shall be released if the right holder fails to pursue the right of action within a reasonable period of time.

Article 54
Notice of Suspension

The importer and the applicant shall be promptly notified of the suspension of the release of goods according to Article 51.

Article 55
Duration of Suspension

If, within a period not exceeding 10 working days after the applicant has been served notice of the suspension, the customs authorities have not been informed that proceedings leading to a decision on the merits of the case have been initiated by a party other than the defendant, or that the duly empowered authority has taken provisional measures prolonging the suspension of the release of the goods, the goods shall be released, provided that all other conditions for importation or exportation have been complied with; in appropriate cases, this time-limit may be extended by another 10 working days. If proceedings leading to a decision on the merits of the case have been initiated, a review, including a right to be heard, shall take place upon request of the defendant with a view to deciding, within a reasonable period, whether these measures shall be modified, revoked or confirmed. Notwithstanding the above, where the suspension of the release of goods is carried out or continued in accordance with a provisional judicial measure, the provisions of paragraph 6 of Article 50 shall apply.

Article 56
Indemnification of the Importer and of the Owner of the Goods

Relevant authorities shall have the authority to order the applicant to pay the importer, the consignee and the owner of the goods appropriate compensation for any injury caused to them through the wrongful detention of goods or through the detention of goods released pursuant to Article 55.

Article 57
Right of Inspection and Information

Without prejudice to the protection of confidential information, Members shall provide the competent authorities the authority to give the right holder sufficient opportunity to have any goods detained by the customs authorities inspected in order to substantiate the right holder's claims. The competent authorities shall also have authority to give the importer an equivalent opportunity to have any such goods inspected. Where a positive determination has been made on the merits of a case, Members may provide the competent authorities the authority to inform the right holder of the names and addresses of the consignor, the importer and the consignee and of the quantity of the goods in question.

Article 58
Ex Officio Action

Where Members require competent authorities to act upon their own initiative and to suspend the release of goods in respect of which they have acquired *prima facie* evidence that an intellectual property right is being infringed:
(a) the competent authorities may at any time seek from the right holder any information that may assist them to exercise these powers;
(b) the importer and the right holder shall be promptly notified of the suspension. Where the importer has lodged an appeal against the suspension with the competent authorities, the suspension shall be subject to the conditions, *mutatis mutandis*, set out at Article 55;
(c) Members shall only exempt both public authorities and officials from liability to appropriate remedial measures where actions are taken or intended in good faith.

Article 59
Remedies

Without prejudice to other rights of action open to the right holder and subject to the right of the defendant to seek review by a judicial authority, competent authorities shall have the authority to order the destruction or disposal of infringing goods in accordance with the principles set out in Article 46. In regard to counterfeit trademark goods, the authorities shall not allow the re-exportation of the infringing goods in an unaltered state or subject them to a different customs procedure, other than in exceptional circumstances.

Article 60
De Minimis Imports

Members may exclude from the application of the above provisions small quantities of goods of a non-commercial nature contained in travellers' personal luggage or sent in small consignments.

SECTION 5: CRIMINAL PROCEDURES

Article 61

Members shall provide for criminal procedures and penalties to be applied at least in cases of wilful trademark counterfeiting or copyright piracy on a commercial scale. Remedies available shall include imprisonment and/or monetary fines sufficient to provide a deterrent, consistently with the level of penalties applied for crimes of a corresponding gravity. In appropriate cases, remedies available shall also include the seizure, forfeiture and destruction of the infringing goods and of any materials and implements the predominant use of which has been in the commission of the offence. Members may provide for criminal procedures and penalties to be applied in other cases of infringement of intellectual property rights, in particular where they are committed wilfully and on a commercial scale.

PART IV
ACQUISITION AND MAINTENANCE OF INTELLECTUAL PROPERTY RIGHTS AND RELATED *INTER-PARTES* PROCEDURES

Article 62

1. Members may require, as a condition of the acquisition or maintenance of the intellectual property rights provided for under Sections 2 through 6 of Part II, compliance with reasonable procedures and formalities. Such procedures and formalities shall be consistent with the provisions of this Agreement.

2. Where the acquisition of an intellectual property right is subject to the right being granted or registered, Members shall ensure that the procedures for grant or registration, subject to compliance with the substantive conditions for acquisition of the right, permit the granting or registration of the right within a reasonable period of time so as to avoid unwarranted curtailment of the period of protection.

3. Article 4 of the Paris Convention (1967) shall apply *mutatis mutandis* to service marks.

4. Procedures concerning the acquisition or maintenance of intellectual property rights and, where a Member's law provides for such procedures, administrative revocation and *inter partes* procedures such as opposition, revocation and cancellation, shall be governed by the general principles set out in paragraphs 2 and 3 of Article 41.

5. Final administrative decisions in any of the procedures referred to under paragraph 4 shall be subject to review by a judicial or quasi-judicial authority. However, there shall be no obligation to provide an opportunity for such review of decisions in cases of unsuccessful opposition or administrative revocation, provided that the grounds for such procedures can be the subject of invalidation procedures.

PART V
DISPUTE PREVENTION AND SETTLEMENT

Article 63
Transparency

1. Laws and regulations, and final judicial decisions and administrative rulings of general application, made effective by a Member pertaining to the subject matter of this Agreement (the availability, scope, acquisition, enforcement and prevention of the abuse of intellectual property rights) shall be published, or where such publication is not practicable made publicly available, in a national language, in such a manner as to enable governments and right holders to become acquainted with them. Agreements concerning the subject matter of this Agreement which are in force between the government or a governmental agency of a Member and the government or a governmental agency of another Member shall also be published.

2. Members shall notify the laws and regulations referred to in paragraph 1 to the Council for TRIPS in order to assist that Council in its review of the operation of this Agreement. The Council shall attempt to minimize the burden on Members in carrying out this obligation and may decide to waive the obligation to notify such laws and regulations directly to the Council if consultations with WIPO on the establishment of a common register containing these laws and regulations are successful. The Council shall also consider in this connection any action required regarding notifications pursuant to the obligations under this Agreement stemming from the provisions of Article 6*ter* of the Paris Convention (1967).

3. Each Member shall be prepared to supply, in response to a written request from another Member, information of the sort referred to in paragraph 1. A Member, having reason to believe that a specific judicial decision or administrative ruling or bilateral agreement in the area of intellectual property rights affects its rights under this Agreement, may also request in writing to be given access to or be informed in sufficient detail of such specific judicial decisions or administrative rulings or bilateral agreements.

4. Nothing in paragraphs 1, 2 and 3 shall require Members to disclose confidential information which would impede law enforcement or otherwise be contrary to the public interest or would prejudice the legitimate commercial interests of particular enterprises, public or private.

Article 64
Dispute Settlement

1. The provisions of Articles XXII and XXIII of GATT 1994 as elaborated and applied by the Dispute Settlement Understanding shall apply to consultations and the settlement of disputes under this Agreement except as otherwise specifically provided herein.

2. Subparagraphs 1(b) and 1(c) of Article XXIII of GATT 1994 shall not apply to the settlement of disputes under this Agreement for a period of five years from the date of entry into force of the WTO Agreement.

3. During the time period referred to in paragraph 2, the Council for TRIPS shall examine the scope and modalities for complaints of the type provided for under subparagraphs 1(b) and 1(c) of Article XXIII of GATT 1994 made pursuant to this Agreement, and submit its recommendations to the Ministerial Conference for approval. Any decision of the Ministerial Conference to approve such recommendations or to extend the period in paragraph 2 shall be made only by consensus, and approved recommendations shall be effective for all Members without further formal acceptance process.

PART VI
TRANSITIONAL ARRANGEMENTS

Article 65
Transitional Arrangements

1. Subject to the provisions of paragraphs 2, 3 and 4, no Member shall be obliged to apply the provisions of this Agreement before the expiry of

a general period of one year following the date of entry into force of the WTO Agreement.

2. A developing country Member is entitled to delay for a further period of four years the date of application, as defined in paragraph 1, of the provisions of this Agreement other than Articles 3, 4 and 5.

3. Any other Member which is in the process of transformation from a centrally-planned into a market, free-enterprise economy and which is undertaking structural reform of its intellectual property system and facing special problems in the preparation and implementation of intellectual property laws and regulations, may also benefit from a period of delay as foreseen in paragraph 2.

4. To the extent that a developing country Member is obliged by this Agreement to extend product patent protection to areas of technology not so protectable in its territory on the general date of application of this Agreement for that Member, as defined in paragraph 2, it may delay the application of the provisions on product patents of Section 5 of Part II to such areas of technology for an additional period of five years.

5. A Member availing itself of a transitional period under paragraphs 1, 2, 3 or 4 shall ensure that any changes in its laws, regulations and practice made during that period do not result in a lesser degree of consistency with the provisions of this Agreement.

Article 66
Least-Developed Country Members

1. In view of the special needs and requirements of least-developed country Members, their economic, financial and administrative constraints, and their need for flexibility to create a viable technological base, such Members shall not be required to apply the provisions of this Agreement, other than Articles 3, 4 and 5, for a period of 10 years from the date of application as defined under paragraph 1 of Article 65. The Council for TRIPS shall, upon duly motivated request by a least-developed country Member, accord extensions of this period.

2. Developed country Members shall provide incentives to enterprises and institutions in their territories for the purpose of promoting and encouraging technology transfer to least-developed country Members in order to enable them to create a sound and viable technological base.

Article 67
Technical Cooperation

In order to facilitate the implementation of this Agreement, developed country Members shall provide, on request and on mutually agreed terms and conditions, technical and financial cooperation in favour of developing and least-developed country Members. Such cooperation shall include assistance in the preparation of laws and regulations on the protection and enforcement of intellectual property rights as well as on the prevention of their abuse, and shall include support regarding the establishment or reinforcement of domestic offices and agencies relevant to these matters, including the training of personnel.

PART VII
INSTITUTIONAL ARRANGEMENTS; FINAL PROVISIONS

Article 68
Council for Trade-Related Aspects of Intellectual Property Rights

The Council for TRIPS shall monitor the operation of this Agreement and, in particular, Members' compliance with their obligations hereunder, and shall afford Members the opportunity of consulting on matters relating to the trade-related aspects of intellectual property rights. It shall carry out such other responsibilities as assigned to it by the Members, and it shall, in particular, provide any assistance requested by them in the context of dispute settlement procedures. In carrying out its functions, the Council for TRIPS may consult with and seek information from any source it deems appropriate. In consultation with WIPO, the Council shall seek to establish, within one year of its first meeting, appropriate arrangements for cooperation with bodies of that Organization.

Article 69
International Cooperation

Members agree to cooperate with each other with a view to eliminating international trade in goods infringing intellectual property rights. For this purpose, they shall establish and notify contact points in their administrations and be ready to exchange information on trade in infringing goods. They shall, in particular, promote the exchange of information and cooperation between customs authorities with regard to trade in counterfeit trademark goods and pirated copyright goods.

Article 70
Protection of Existing Subject Matter

1. This Agreement does not give rise to obligations in respect of acts which occurred before the date of application of the Agreement for the Member in question.

2. Except as otherwise provided for in this Agreement, this Agreement gives rise to obligations in respect of all subject matter existing at the date of application of this Agreement for the Member in question, and which is protected in that Member on the said date, or which meets or comes subsequently to meet the criteria for protection under the terms of this Agreement. In respect of this paragraph and paragraphs 3 and 4, copyright obligations with respect to existing works shall be solely determined under Article 18 of the Berne Convention (1971), and obligations with respect to the rights of producers of phonograms and performers in existing phonograms shall be determined solely under Article 18 of the Berne Convention (1971) as made applicable under paragraph 6 of Article 14 of this Agreement.

3. There shall be no obligation to restore protection to subject matter which on the date of application of this Agreement for the Member in question has fallen into the public domain.

4. In respect of any acts in respect of specific objects embodying protected subject matter which become infringing under the terms of legislation in conformity with this Agreement, and which were commenced, or in respect of which a significant investment was made, before the date of acceptance of the WTO Agreement by that Member, any Member may provide for a limitation of the remedies available to the right holder as to the continued performance of such acts after the date of application of this Agreement for that Member. In such cases the Member shall, however, at least provide for the payment of equitable remuneration.

5. A Member is not obliged to apply the provisions of Article 11 and of paragraph 4 of Article 14 with respect to originals or copies purchased prior to the date of application of this Agreement for that Member.

6. Members shall not be required to apply Article 31, or the requirement in paragraph 1 of Article 27 that patent rights shall be enjoyable without discrimination as to the field of technology, to use without the authorization of the right holder where authorization for such use was granted by the government before the date this Agreement became known.

7. In the case of intellectual property rights for which protection is conditional upon registration, applications for protection which are pending on the date of application of this Agreement for the Member in question shall be permitted to be amended to claim any enhanced protection provided under the provisions of this Agreement. Such amendments shall not include new matter.

8. Where a Member does not make available as of the date of entry into force of the WTO Agreement patent protection for pharmaceutical and

agricultural chemical products commensurate with its obligations under Article 27, that Member shall:

(a) notwithstanding the provisions of Part VI, provide as from the date of entry into force of the WTO Agreement a means by which applications for patents for such inventions can be filed;
(b) apply to these applications, as of the date of application of this Agreement, the criteria for patentability as laid down in this Agreement as if those criteria were being applied on the date of filing in that Member or, where priority is available and claimed, the priority date of the application; and
(c) provide patent protection in accordance with this Agreement as from the grant of the patent and for the remainder of the patent term, counted from the filing date in accordance with Article 33 of this Agreement, for those of these applications that meet the criteria for protection referred to in subparagraph (b).

9. Where a product is the subject of a patent application in a Member in accordance with paragraph 8(a), exclusive marketing rights shall be granted, notwithstanding the provisions of Part VI, for a period of five years after obtaining marketing approval in that Member or until a product patent is granted or rejected in that Member, whichever period is shorter, provided that, subsequent to the entry into force of the WTO Agreement, a patent application has been filed and a patent granted for that product in another Member and marketing approval obtained in such other Member.

Article 71
Review and Amendment

1. The Council for TRIPS shall review the implementation of this Agreement after the expiration of the transitional period referred to in paragraph 2 of Article 65. The Council shall, having regard to the experience gained in its implementation, review it two years after that date, and at identical intervals thereafter. The Council may also undertake reviews in the light of any relevant new developments which might warrant modification or amendment of this Agreement.

2. Amendments merely serving the purpose of adjusting to higher levels of protection of intellectual property rights achieved, and in force, in other multilateral agreements and accepted under those agreements by all Members of the WTO may be referred to the Ministerial Conference for action in accordance with paragraph 6 of Article X of the WTO Agreement on the basis of a consensus proposal from the Council for TRIPS.

Article 72
Reservations

Reservations may not be entered in respect of any of the provisions of this Agreement without the consent of the other Members.

Article 73
Security Exceptions

Nothing in this Agreement shall be construed:

(a) to require a Member to furnish any information the disclosure of which it considers contrary to its essential security interests; or

(b) to prevent a Member from taking any action which it considers necessary for the protection of its essential security interests;

(i) relating to fissionable materials or the materials from which they are derived;

(ii) relating to the traffic in arms, ammunition and implements of war and to such traffic in other goods and materials as is carried on directly or indirectly for the purpose of supplying a military establishment;

(iii) taken in time of war or other emergency in international relations; or

(c) to prevent a Member from taking any action in pursuance of its obligations under the United Nations Charter for the maintenance of international peace and security.

References

Akerlof, George A. (1970) "The Market for Lemons: Quality Uncertainty and the Market Mechanism", *Quarterly Journal of Economics* 84(3, Aug.): 488–500.

Akerele, Olayiwola (1990) "Medicinal Plants in Traditional Medicine", in H. Wagner and Norman R. Farnsworth, eds., *Economic and Medicinal Plant Research: Plants in Traditional Research*, Vol. 4, London/Boston/New York: Academic Press, pp. 5–16.

Alexiades, Miguel N. (2002) "Cat's Claw", in Patricia Shanley, Alan Pierce, Sarah A. Laird and Abraham Guillen, *Tapping the Green Market, Certification and Management of Non-Timber Forest Products*, London and Sterling, VA: Earthscan Publications, pp. 93–109.

Anderson, Edward F. (1986a) "Ethnobotany of Hill Tribes of Northern Thailand. I. Medicinal Plants of Akha", *Economic Botany* 40(1): 38–53.

——— (1986b) "Ethnobotany of Hill Tribes in Northern Thailand. II. Lahu Medicinal Plants", *Economic Botany* 40(4): 442–450.

Angell, Marcia (2004) *The Truth About Drug Companies*, New York: Random House.

Archibugi, Daniele and Michie, Jonathan (1995) "The Globalisation of Technology: A New Taxonomy", *Cambridge Journal of Economics* 19: 121–140.

Arora, Ashish (1995) "Appropriating rents from Innovation: A Historical Look at the Chemical Industry", in Horst Albach and Stephanie Rosenkranz, eds., *Intellectual Property Rights and Global Competition: Towards a New Synthesis*, Berlin: WZB Publications, pp. 339–368.

Arrow, Kenneth J. (1962) "Economic Welfare and the Allocation of Resources for Invention", in National Bureau of Economic Research, *The Rate and*

Direction of Inventive Activity, Princeton: Princeton University Press, pp. 609–625,

——— (1969) "The Organization of Economic Activity: Issues Pertinent to the Choice of Market versus Non-market Allocation", in *The Analysis and Evolution of Public Expenditure: The PPB system* 1 US Joint Economic Committee, 91st Congress, 1st Session, Washington DC: US Government Printing Office, pp. 59–73.

Arrow, Kenneth J. and Fisher, A.C. (1974) "Environmental Preservation, Uncertainty and Irreversibility", *Quarterly Journal of Economics* 88(2): 312–319.

Artuso, Anthony A. (1997) "Capturing the Chemical Value of Biodiversity: Economic Perspectives and Policy Prescriptions", in Francesca T. Grifo and Joshua Rosenthal, eds., *Biodiversity and Human Health*, Covelo CA: Island Press, pp. 187–204.

——— (1997a) *Drugs of Natural Origin: Economic and Policy Aspects of Discovery, Development and Marketing*, New York/London: Pharmaceutical Products Press.

——— (1997b) "Capturing the Chemical Value of Biodiversity: Economic Perspectives and Policy Prescriptions", in Francesca T. Grifo and Joshua Rosenthal, eds., *Biodiversity and Human Health*, Washington DC: Island Press, pp. 187–204.

Ayensu, Edward S. (1983) "Endangered Plants Used in Traditional Medicine", in Robert H. Bannermann, John Burton, and Ch'en Wen-Chieh, eds., *Traditional Medicine and Health Care Coverage: Reader for Health Administrators and Practitioners*, Geneva: WHO, pp. 175–280.

Aylward, Bruce A. (1993) "The Economic Value of Pharmaceutical Prospecting and its Role in Biodiversity Conservation", LEEC Discussion Paper No. 93-05, London: IEDD.

——— (1995) "The Role of Plant Screening and Plant Supply in Biodiversity Conservation, Drug Development and Health Care", in Timothy M. Swanson, ed., *Intellectual Property Rights and Biodiversity Conservation: An Interdisciplinary Analysis of the Values of Medicinal Plants*, Cambridge: Cambridge University Press, pp. 93–127.

——— (1996) "Capturing the Pharmaceutical Value of Species Information: Opportunities for Developing Countries", in Michael J. Balick, Elaine Elisabetsky, and Sarah A. Laird, eds., *Medicinal Resources of the Tropical Forest: Biodiversity and its Importance to Human Health*, New York: Columbia University Press, pp. 219–229.

Balandrin, M., Klocke, J., Wurtele, F.E., Bollinger, W. (1985) "Natural Plant Chemicals: Sources of Industrial and Medicinal Materials", *Science* 288: 1154–1160.

Balick, Michael J. (1990) "Ethnobotany and the Identification of Therapeutic Agents from the Rainforest", in Derek J. Chadwick and Joan Marsh, *Bioactive compounds from Plants*, UBA Foundation Symposium, New York: John Wiley, pp. 26–28.

Balick, Michael J., Elisabetsky, Elaine, and Laird, Sarah A. (1996) *Medicinal Resources of the Tropical Rainforest: Biodiversity and its Importance to Human Health*, New York: Colombia University Press.

Ballance, Robert, Pogány, János, and Forstner, Helmut (1992) *The World's Pharmaceutical Industries: An International Perspective on Innovation, Competition and Policy*, Aldershot and Vermont: Edward Elgar, prepared for the United Nations Industrial Development Organization.

Bandhopadhyay, A.K. and Saha, G.S. (1998) "Indigenous Methods of Seed Selection and Preservation on the Andaman Islands in India", *Indigenous Knowledge and Development Monitor* 6(1): 3–6.

Bell, J. (1997) "Biopiracy's Latest Disguises", in *Seedling*, Quarterly Newsletter of GRAIN (Genetic Resources Action International) 14(2, June).

Berlin, Brent, Berlin, Elois Ann, Fernández Ugalde, et al. (1999) "The Maya ICBG: Drug Discovery, Medical Ethnobiology and Alternative Forms of Economics Development in the Highland Maya Region of Chiapas, Mexico", *Pharmaceutical Biology* 37(Supplement): 127–144.

Berlin, Elois A. and Berlin, Brent (1999) *Whose Knowledge? Whose Property? The Case of OMIECH, RAFI and the Maya ICBG*, 13 December, available at: http://guallart.dac.uga.edu/ICBGreply.html

Biber-Klemm, Susette (2000) "The Protection of Traditional Knowledge on the International Level – Reflections in Connection with World Trade", paper submitted to UNCTAD Expert Meeting on Systems and National Experiences for Protecting Traditional Knowledge, Innovations and Practices, Geneva, 30 October–1 November (on file with author).

Bragdon, Susan H. and Downes, David R. (1998) "Recent Policy Trends and Developments Related to the Conservation, Use and Development of Genetic Resources", *Issues in Genetic Resources*: i–42.

Braithwaite, John and Drahos, Peter F. (2000) *Global Business Regulation*, Cambridge: Cambridge University Press.

Bromley, Daniel W. and Cyrene, Michael M. (1989) "The Management of Common Property Natural Resources: Some Common Conceptual and Operational Fallacies", World Bank Discussion Papers No 57, Washington DC: World Bank.

Brown, Gardner M. (2000) "Renewable Natural Resource: Management and Use without Markets", *Journal of Economic Literature* 38(4): 875–914.

Burhenne-Guilmin, Françoise and Glowka, Lyle (1994) "An Introduction to the Convention on Biological Diversity", in A.F. Krattinger, J.A. McNeely, W.H. Lesser, et al., eds., *Widening Perspectives on Biodiversity*, Geneva: The World Conservation Union (IUCN), pp. 15–18.

Calandrillo, Steve P. (1998) "An Economic Analysis of Property Rights in Information: Justifications and Problems of Exclusive Rights, Incentives to Generate Information, and the Alternative of a Government-Run Reward System", *Fordham Intellectual Property, Media and Entertainment Law Journal* IX(1): 301–359.

CEAS Consultants, Tansey, Geoff and Queen Mary Intellectual Property Research Institute (2000) *Study on the Relationship between the Agreement on TRIPS and Biodiversity Related Issues: Final Report for DG Trade European Commission*, September 2000.

Chapela, Ignacio H. (1997) "Using Fungi from a Node of Biodiversity: Conservation and Property Rights in Oaxacan Forests", in K.E. Hoagland and

A.Y. Rossman, eds., *Global Genetic Resources: Access, Ownership and Intellectual Property Rights*, USA: Association of Systematic Collections, pp. 165–180.

Coase, Ronald H. (1937) "The Nature of the Firm", *Economica* 4(15): 386–405.

——— (1960) "The Problem of Social Cost", *Journal of Law and Economics* 3(1): 1–44.

Correa, Carlos M. (1998) "Patent Rights," in Carlos M. Correa and Abdulqawi A. Yusuf, eds., *Intellectual Property Rights and Internatinal Trade: the TRIPS Agreement*, London: Kluwer Law International, pp. 189–221.

——— (2000) *Implementing the TRIPS Agreement in the Patent Field*, London: Zed Books.

——— (2001) "Traditional Knowledge and Intellectual Property Issues and Options Surrounding the Protection of Traditional Knowledge", Discussion Paper, Geneva: Quaker United Nations Office (QUNO).

——— (2003) "The Access Regime and the Implementation of the FAO International Treaty on Plant Genetic Resources for Food and Agriculture in the Andean Group Countries", *Journal of World Intellectual Property* 6(6): 795–806.

——— (2004) "Intellectual Property after Doha: Can Developing Countries Move Forward Their Agenda on Biodiversity and Traditional Knowledge?" Technology Policy Brief, Vol. 3, No. 2, Maastricht: United Nations University Institute for New Technologies (UNU-INTECH).

Cottier, Thomas (1991) "The Prospects for Intellectual Property in the GATT", *CML Review* 28: 383–414.

Coughlin Jr., M.D. (1993) "Comment: Using the Merck-INBio Agreement to Clarify the Convention on Biological Diversity", *Columbia Journal of Transnational Law* 31(2): 337–375.

Cox, Paul A. (1990) "Ethnopharmacology and the Search for New Drugs", in D.J. Chadwick and Arsch, J. eds., *Bioactive Compounds from Plants*, Cambridge: Cambridge University Press, pp. 40–65.

——— (1995) "Shaman as a Scientist", in Kurt Hostettmann, A. Marston, M. Maillard, and M. Hamburger, eds., *Phytochemistry of Plants used in Traditional Medicine*, Oxford: Clarendon Press, pp. 1–16.

Cox, Paul A., Sperry, L.R., Tuominen, M., and Bohlin, L. (1983) "Pharmacological Activity of the Samoan Ethnopharmacopoeia", *Economic Botany* 43(4): 487–497.

Cragg, Gordon M. and Boyd, Michael R. (1996) "Drug Discovery and Development at the National Cancer Institute: The Role of Natural Products of Plant Origin", in Michael J. Balick, Elaine Elisabetsky, and Sarah A. Laird, eds., *Medicinal Resources of the Tropical Forests: Biodiversity and its Importance to Human Health*, New York: Columbia University Press, pp. 101–136.

Cragg, Gordon M. and Newman, David J. (2004) "Biodiversity: A Countinuing Source of Novel Leads", *Pure and Applied Chemistry*, in press.

Cragg, G.M., Newman, D.J., and Snader, K.M. (1997) "Natural Products in Drug Discovery and Development", *Journal of Natural Products* 60(1): 52–60.

Cragg, Gordon M., Boyd, Michael R., Grever, Michael R., Schepartz, Saul A. (1995) "Pharmaceutical Prospecting and the Potential for Pharmaceutical Crops: Natural Product Drug Discovery and Development at the United States

National Cancer Institute", *Annals of the Missouri Botanical Garden* 82(1): 47–53.

Dalton, Rex (2001) "The Curtain Falls", *Nature* 414(13 December): 685, www.nature.com/nature.

——— (2004a) "Bioprospectors hunt for fair share of profits", *Nature* 427(12 February): 576 (www.nature.com/nature).

——— (2004b) "Bioprospects less than golden", *Nature* 429(10 June): 598–600 (www.nature.com/nature).

Daly, Douglas C. (1992) "The National Cancer Institute's Plant Collections Program: Update and Implications for Tropical Forests", in Marc Plotkin and L. Famolare, eds., *Sustainable Harvest and Marketing of Rain Forest Products*, Washington DC/Covelo: Island Press, pp. 224–230.

Daly, Douglas C. and Limbach, Charles F. (1996) "The Contribution of the Physician to Medicinal Plant Research", in Michael J. Balick, Elaine Elisabetsky and Sarah A. Laird, eds., *Medicinal Resources of the Tropical Forests: Biodiversity and its Importance to Human Health*, New York: Columbia University Press, pp. 48–62.

David, Paul (1993) "Intellectual Property Institutions and the Panda's Thumb: Patents, Copyrights and Trade Secrets in Economic Theory and History", in M.B. Wallerstein, M.E. Mogee, and R.A. Schoen, eds., *Global dimensions of intellectual property rights in science and technology*, Washington, DC: National Academy of Sciences Press, pp. 19–62.

Demsetz, Harold (1964) "Towards a Theory of Property Rights", *Journal of Law and Economics* 7(11): 11–26.

Dhillon, Shivcharn S. and Amundsen, Catherine (2000) "Bioprospecting and the Maintenance of Biodiversity", in Hanne Svarstad and Shivcharn S. Dhillon, eds., *Bioprospecting: From Biodiversity in the South to Medicines in the North*, Oslo: Spartacus Forlag As, pp. 103–132.

Drahos, Peter F. (1995) "Global Property Rights in Information: The Story of TRIPS at the GATT", *Prometheus* 13(1): 6–19.

——— (1996) *A Philosophy of Intellectual Property*, Aldershot: Dartmouth Publishing.

Dutfield, Graham (2000) *Intellectual Property Rights, Trade and Biodiversity*, London: Earthscan.

——— (2003) *Intellectual Property Rights and the Life Science Industries*, Aldershot: Ashgate Publishing.

——— (2004) "Outstanding Issues in the Protection of Traditional Knowledge", *Technology Policy Brief* 3(2) Maastricht: United Nations University Institute for New Technologies (UNU-INTECH).

Eisner, Thomas (1991) "Chemical prospecting: a proposal for action", in F.H. Bormann, and Stephen R. Kellert, eds., *Ecology, Ethics and Economics: The Broken Circle*, New Haven/London: Yale University Press, pp. 196–202.

Elisabetsky, Elaine (1991) "Folklore, Tradition or Know-How?" *Cultural Survival Quarterly*, Summer: 9–13.

Ellickson, Robert (1994) *Order without Law*, Boston: Harvard University Press.

Etkin, Nina L. (2003) "Benefit-Sharing: Hype or Hope?", working paper, International Conference on Medicinal Plants, Access Use and Benefit-Sharing in Light of the CBD, Oslo: University of Oslo, 3 April.

Etkin, N.L. and John, T. (1998) "Pharmafoods and Nutriceuticals: Paradigm Shifts in Biotherapeutics", in H.D.V. Prendergast, N.L. Etkin, D.R. Harris, and P.J. Houghton, eds., *Plants for Food and Medicine*, Kew: Royal Botanical Gardens, pp. 3–16.

European Commission (EC) (2001) "Review of the Provisions of Article 27(3)(b) of the TRIPs Agreement", Communication by the European Communities and their Member States on the Relationship Between the Convention on Biological Diversity and the TRIPs Agreement, a Submission to the WIPO Intergovernmental Committee on Intellectual Property and Genetic Resources, Traditional Knowledge and Folklore, First Session, Geneva, 30 April to 3 May.

Farnsworth, N.R. (1988) "Screening Plants For New Medicines", in E.O. Wilson, ed., *Biodiversity*, Washington DC: National Academy Press, pp. 83–97.

Farnsworth, Norman R. and Soejarto, D.D. (1985) "Potential Consequences of Plant Extinction in the U.S. on the Availability of Prescription Drugs", *Economic Botany* 39(3): 231–240.

Fellows, Linda and Scofield, Anthony (1995) "Chemical Diversity in Plants", in Timothy Swanson, ed., *Intellectual Property Rights and Biodiversity Conservation: An Interdisciplinary Analysis of the Values of Medicinal Plants*, Cambridge: Cambridge University Press, pp. 19–44.

Foray, Dominique (1995) "Knowledge Distribution and the Institutional Infrastructure: The Role of Intellectual Property Rights", in Horst Albach and Stephanie Rosenkranz, eds., *Intellectual Property Rights and Global Competition: Towards a New Synthesis*, Berlin: WZB Publications, pp. 77–118.

Freedman, M.D.A. (1994) *Lloyd's Introduction to Jurisprudence*, 6th edn, London: Sweet and Maxwell.

Furubotn, Eirik G. and Richter, Rudolf (1998) *Institutions and Economic Theory: The Contribution of the New Institutional Economics*, Ann Arbor: University of Michigan Press, pp. 39–67.

Gadgil, Madhav and Devasia, P. (1995) "Intellectual Property Rights and Biological Resources: Specifying Geographical Origins and Prior Knowledge of Uses", *Current science* 69(8): 637–639.

Gallini, Nancy and Scotchmer, Suzanne (2002) "Intellectual Property: When is it the Best Incentive System?", in Adam Jaffe, Joshua Lerner, and Scott Stern, eds., *Innovation Policy and the Economy*, Cambridge Massachusetts: MIT Press, pp. 51–78.

Ganguli, P. (2000) "Intellectual Property Rights: Imperatives for the Knowledge Industry", *World Patent Information* 22: 167–175.

Gehl Sampath, Padmashree (2003) "Designing National Regimes that Promote Public Health", Discussion Paper, 2003–8, Maastricht: United Nations University Institute for New Technologies (UNU-INTECH).

Gehl Sampath, Padmashree and Tarasofsky, Richard G. (2002) *Study on the Inter-Relations between Intellectual Property Rights Regimes and the Conservation of Genetic Resources*, paper presented to European Commission Directorate-General Environment, Final Report, Berlin, 31 December.

Gereffi, G. (1983) *The Pharmaceutical Industry and Dependency in the Third World*, Princeton: Princeton University Press.
Glowka, Lyle (2001) *Towards a Certification System for Bioprospecting Activities*, Bern: Study commissioned by the State Secretariat for Economics Affairs.
Glowka, Lyle, Françoise Burheen-Guilmin, Hugh Synge, et al. (1995) *A Guide to the Convention on Biological Diversity*, Gland: World Conservation Union (IUCN).
Gollin, Michael A. (1993) "An Intellectual Property Rights Framework for Biodiversity Prospecting", in Walter V. Reid, Sarah A. Laird, et al., *Biodiversity Prospecting: Using Genetic Resources for Sustainable Development*, Washington, USA: World Resources Institute Publication, pp. 159–198.
Gopalakrishnan, N.S. (2002) "Protection of Traditional Knowledge: The Need for a *Sui Generis* Law in India", *Journal of Intellectual Property* 5(5): 725–743.
Grenier, Louis (1998) *Working with Indigenous Knowledge: A Guide for Researchers*, Ottawa: International Development Research Centre.
Grifo, F.T. (1996) "Chemical Prospecting: An Overview of the International Cooperative Biodiversity Groups Program", in *Biodiversity, Biotechnology and Sustainable Development in Health and Agriculture*, Washington DC: Pan American Health Organization.
Grossman, S.J. and Hart, Oliver D. (1986) "The Costs and Benefits of Ownership: A Theory of Vertical and Lateral Integration", *Journal of Political Economy* 94(4): 619–719.
Grossmann, Stanford J. and Stiglitz, Joseph (1980) "On the Impossibility of Informationally Efficient Markets", *American Economic Review* 70(3): 393–408.
Haas, Lucian and Schwaegerl, Christian (2002) "Der Untergang der Biopiraten", *Frankfurter Allgemeine Zeitung* 11(April): 50.
Halliday, R.G., Walker S.R., Lumley, C.E. (1992) "R & D Philosophy and Management in the World's Leading Pharmaceutical Companies", *Journal of Pharmaceutical Medicine* 2: 139–154.
Halper, Mark (2004) "Lab Leaders", *Time*, 13 December, pp. 57–60.
Hart, H.L.A. (1955) "Are there any Natural Rights?", *Philosophical Review* 64: 175.
Hart, Oliver D. (1989) "An Economist's Perspective on the Theory of the Firm", *Columbia Law Review* 89(7): 1757–1774.
——— (1995) *Firms, Contracts and Financial Structure*, Oxford: Clarendon.
Hart, Oliver D. and Moore, John (1990) "Property Rights and the Nature of the Firm", *Journal of Political Economy* 98(6): 1119–1158.
——— (1998) "Foundations of Incomplete Contracts", NBER Working Paper No. 6726, National Bureau of Economic Research, September.
Hayden, Cori (2003) *When Nature Goes Public: The Making and Unmaking of Bioprospecting in Mexico*, Princeton: Princeton University Press.
Henry, C. (1974) "Investment Decisions under Uncertainty: The 'Irreversibility Effect'", *American Economic Review* 64(6): 1006–1112.
Hettinger, Edwin C. (1989) "Justifying Intellectual Property", Philosophy and Public Affairs, Vol. 31, in Peter F. Drahos, ed. (1999) *Intellectual Property*, Aldershot/Vermont: Dartmouth Publishing/Ashgate Publishing, pp. 31–52.

Hirschleifer, Jack and Riley, John G. (1995) *The Economics of Uncertainty and Information*, Cambridge: Cambridge University Press.

Holmstroem, Bengt and Milgrom, Paul (1991) "Multi-Task Principal-Agent Analyses: Incentive Contracts, Asset Ownership and Job Design", *Journal of Law, Economics and Organization* 7 (Special Issue): 24–52.

Holmström, Bengt and Roberts John (1998) "The Boundaries of the Firm Revisited", *Journal of Economic Perspectives* 12(4): 73–94.

Huft, Michael J. (1995) "Indigenous Peoples and Drug Discovery Research: A Question of Intellectual Property Rights", *Northwestern University Law Review* 89(4): 1652–1730.

IMS Retail Drug Monitor (2004a) "Tracking 13 Key Global Pharma Markets", IMS Health, available at www.imshealth.com.

——— (2004b) "Biogenerics: Waiting for the Green Light", IMS Health, available at: www.imshealth.com.

Jaffe, Adam B. (1999) "The US Patent System in Transition: Policy Innovation and the Innovation Process", Working Paper 7280, Cambridge MA: National Bureau of Economic Research (http://www.nber.org/papers/w7280).

Kanbur, Ravi (1992) "Heterogeneity, Distribution and Cooperation in Common Property Resource Management", background paper for 1992 World Development Report, Policy Research Working Papers, World Bank, January, WPS 844.

Katz, Michael L. (1989) "Vertical Contractual Relationships", in R. Schmalensee and R.D. Willig, eds., *The Handbook of Industrial Organization*, Amsterdam: North-Holland, pp. 655–721.

Katz, Michael L. and Rosen, Harvey S. (1998) *Microeconomics*, 3rd Edn, Boston, MA: Irwin, McGraw Hill.

Kingston, David G., Abdel-Kader, Maged, Zhou, Bing-Nan, et al. (1999) "The Suriname International Cooperative Biodiversity Group Program: Lessons from the First Five Years", *Pharmaceutical Biology* 37(Supplement): 22–34.

Klein, Benjamin and Leffler, K.B. (1981) "The Role of Market Forces in Assuring Contractual Performance", *Journal of Political Economy* 89(4): 615–641.

Klein, Benjamin, Crawford, Robert G., and Alchian, Armen (1978) "Vertical Integration, Appropriable Rents and the Competitive Contracting Process", *Journal of Law and Economics* 21(2): 297–326.

Kothari, A. and Anuradha, R.V. (1997) "Biodiversity, intellectual property rights, and the GATT agreement: how to address the conflicts?", *Biopolicy* 2(4): PY97004, online journal (http://www.bdt.org.br/bioline/py).

Kothari, Ashish and Das, Priya (2000) "Local Community Knowledge and Practices in India", in Darrell A. Posey and Oxford Centre for the Environment, Ethics and Society, eds., *Cultural and Spiritual Values of Biodiversity*, United Nations Environment Programme (UNEP), Intermediate Technology Publications, pp. 190–194.

Krasser, R. (1996) "The Protection of Trade Secrets in the TRIPS Agreement", in Karl Brier and Gerhard Shricker, eds., *From GATT to TRIPS: The Agreement on Trade Related Aspects of Intellectual Property Rights*, IIC Studies 18, Munich: Max Planck Institute for Patent Copyright and Competition Law, pp. 216–225.

Krattiger, Anatole F. and Lesser, William H. (1995) "The 'Facilitator': Proposing a New Mechanism to Strengthen the Equitable and Sustainable Use of Biodiversity", *Environmental Conservation* 22(3): 211–215.

Kreps, David M. (1995) *A Course in Microeconomic Theory*, Princeton, New Jersey: Princeton University Press.

Laird, Sarah A. (1993) "Contracts for Biodiversity Prospecting", in Walter V. Reid, Sarah A. Laird, R. Gamez, et al., *Biodiversity Prospecting: Using Genetic Resources for Sustainable Development*, Washington, DC: World Resources Institute, National Biodiversity Institute of Costa Rica, Rainforest Alliance, and African Centre for Technology, pp. 99–130.

Laird, Sarah A. and Guillén, Abraham (2002) "Marketing Issues", in Patricia Shanley, Alan Robert Pierce, Sarah A. Laird, and S. Abraham Guillén, eds., *Tapping the Green Market*, London/Sterling: Earthscan Publications, pp. 322–336.

Laird, Sarah A. and Lisinge, Esterine (1998) "Benefit-Sharing Case Studies: *Aristocladus korupensis* and *Prunus africana*", UNEP/CBD/COP/4/Inf.25, available at http://www.biodiv.org/doc/case-studies/abs/cs-abs-aristo.pdf

——— (2002) "Protected area research policies: developing a basis for equity and accountability", in Sarah A. Laird, ed., *Biodiversity and Traditional Knowledge, Equitable Partnerships in Practice*, Earthscan Publications, London, pp. 127–176.

Laird, Sarah A. and Pierce, Alan R. (2002) "Promoting Sustainable and Ethical Botanicals: Strategies to Improve Commercial Raw Material Sourcing", result from the Sustainable Botanicals Pilot Project Industry Surveys, Case Studies and Standards Collection, Final Report, New York, May.

Laird, Sarah A. and ten Kate, Kerry (2002) "Biodiversity prospecting: the commercial use of genetic resources and best practice in benefit-sharing", in Sarah A. Laird, ed., *Biodiversity and Traditional Knowledge: Equitable Partnerships in Practice*, London/Sterling: Earthscan Publications.

Laird, Sarah A., Alexiades, Miguel N., Bannister, Kelly P., and Posey, Darrell A. (2002) "Publication of Biodiversity Research Results and the Flow of Knowledge", in Sarah A. Laird, ed., *Biodiversity and Traditional Knowledge*, London: Earthscan Publications, pp. 77–101.

Laird, Sarah A., ten Kate, Kerry, Fellows, Linda, and Scofield, Anthony (1995) "Chemical Diversity in Plants", in Timothy M. Swanson, ed., *Intellectual Property Rights and Biodiversity Conservation: An Interdisciplinary Analysis of the Values of Medicinal Plants*, Cambridge: Cambridge University Press, pp. 19–44.

Landes, William M. and Posner, Richard (1989) "An Economic Analysis of Copyright Law", *Journal of Legal Studies* 18(2): 325–365.

Lange, D. (1997) "Trade in Plant Material for Medicinal and Other Purposes: A German Case Study", *TRAFFIC Bulletin* 17(1): 21–32.

Lassere, Pierre (2002) "Evaluation of Biodiversity: A Real Options Approach", paper presented at a Maastricht Economic Research Institute on Innovation and Technology (MERIT) Seminar, Maastricht, 3 December.

Lazaroff, C. (2000) "Forests Plants Help Battle Tuberculosis", *Environmental News Service*, 4 August.

Lesser, William H. (1995) *Equitable Patent Protection in the Developing World*, Christchurch: Eubios Ethics Institute.
——— (1996) "Using Access Legislation to Provide Intellectual Property Rights Protection for Indigenous Knowledge", Draft 4/30, on file with the author.
Lettington, Robert J.L. and Nnadozie, Kent (2003) "A Review of the Intergovernmental Committee on Genetic Resources, Traditional Knowledge and Folklore at WIPO", Trade-Related Agenda, Development and Equity, Occasional Paper, Geneva: World Intellectual Property Organization.
Lewis, D.M. (1992) *Millennium: Tribal Wisdom and the Modern World*, New York: Viking Publications.
Lewis, Walter H., Lamas, G., Vaisberg, A., Corley, D.G., and Sarasara, C. (1999) "Peruvian Medicinal Plant Sources of New Pharmaceuticals (International Cooperative Biodiversity Group)", *Pharmaceutical Biology* 37(Supplement): 69–83.
Lightbourne, Muriel (2003) "Of Rice and Men an Attempt to Assess the Basmati Affair", *Journal of World Intellectual Property* 6(6): 875–894.
McCay, B.B. (1995) "Social and Ecological Implications of ITQs; An Overview", *Ocean and Coastal Management* 28(1–3): 3–22.
Mandeville, Thomas (1999) "An Information Economics Perspective on Information," in Peter F. Drahos, ed., *Intellectual Property*, Aldershot/Vermont: Dartmouth Publishing/Ashgate Publishing.
Mann, Howard J. (1996) "Intellectual Property Rights, Biotechnology and the Protection of Biological Diversity: Literature Review", prepared for Industry Canada, Ottawa: Intellectual Property Policy Directorate.
Marguiles, Rebecca (1993) "Protecting Biodiversity: Recognizing Intellectual Property Rights in Plant Genetic Resources," *Michigan Journal of International Law* 14: 322–356.
Mas-Colell, Andreu, Winston, Michael D. and Green, Jerry R. (1995) *Microeconomic Theory*, Oxford: Oxford University Press.
May, Christopher (2000) *A Global Political Economy of Intellectual Property Rights: The New Enclosures?*, London: Routledge Publishers.
McCaleb, Robert S. (1994) "Medicinal Plants for Healing the Planet: Biodiversity and Environmental Health Care", in F. Grifo and J. Rosenthal, eds., *Biodiversity and Human Health*, Washington DC: Island Press, pp. 221–242.
McChesney, P. (1992) "Biological Diversity; Chemical Diversity and the Search For New Pharmaceuticals", Tropical Forest Medical Resources and the Conservation of Biodiversity, presentation at the Rainforest Alliance Symposium, 1992.
McChesney, J.D. (1996) "Biological Diversity, Chemical Diversity and the Search for New Pharmaceuticals", in Michael J. Balick, Elaine Elisabetsky, and Sarah, A. Laird, eds., *Medicinal Resources of the Tropical Rainforest: Biodiversity and its Importance to Human Health*, New York: Colombia University Press.
McNeely, Jeffrey A. (2000) "Economics and Conserving Forest Genetic Diversity", in T. Boyle, A. Young, and D. Boshier, eds., *Forest Conservation Genetics: Principles and Practice*, Gland: World Conservation Union (IUCN), http://economics.iucn.org.

McNeely, Jeffrey and Vorhies, Frank (2000) "Economics and Conserving Forest Genetic Diversity", in T. Boyle, A. Young, and D. Boshier, eds., *Forest Conservation Genetics: Principles on Early Experiences Around the World*, Gland: World Conservation Union (IUCN), http://economics.iucn.org.

Mendelsohn, Robert and Balick, Michael J. (1995) "The Value of Undiscovered Pharmaceuticals in Tropical Forests", *Economic Botany* 49(2): 223–228.

Milgrom, Paul and Roberts, John (1986) "Price and Advertising Signals of Product Quality", *Journal of Political Economy* 94(4): 796–821.

Milgrom, Paul J. and Roberts, John D. (1996) *Economics, Organization and Management*, New Jersey: Prentice Hall.

Miller, Henry I. (1997) *Policy Controversy in Biotechnology: an Insider's View*, Georgetown, Landes Bioscience Publication.

Miller, Kenton R. and Lanou, Steven M. (1995) *National Biodiversity Planning: Guidelines Based on Early Experiences around the World*, Gland: World Resources Institute, United National Environment Programme and World Conservation Union.

Mooney, Pat (2000) "Why We Call It Biopiracy", in Hanne Svarstad and Shivcharn S. Dhillon, eds., *Bioprospecting: From Biodiversity in the South to Medicines in the North*, Oslo: Spartacus Forlag.

Mountford, Helen and Keppler, Jan Horst (1999) "Financing Incentives for the Protection of Biodiversity", *The Science of the Total Environment* 240(1–3): 91–105.

Mundy, Paul (1993) "Indigenous Communication Systems: An Overview", *Journal of SID* 3: 41–44.

Myerson, R.B. (1981) "Optimal Auction Design", *Mathematics of Operations Research* 6(1): 58–73.

——— (1991) *Game Theory: Analysis of Conflict*, Cambridge: Harvard University Press.

Nelson, Philip (1974) "Advertising as Information", *Journal of Political Economy* 82(4): 729–754.

Newman, David J. and Laird, Sarah A. (1999) "The Influence of Natural Products on 1997 Pharmaceutical Sales Figures", in Kerry ten Kate and Sarah A. Laird, eds., *The Commercial Use of Biodiversity*, London: Earthscan Publications, pp. 333–335.

Newman, David. J. and Cragg, Gordon M. (2004) "Advanced Preclinical and Clinical Trials of Natural Products and Related Compounds from Marine Sources", *Journal of Natural Products* 67: 1216–1238.

Newman, David J., Cragg, Gordon M., and Snader, Kenneth M. (2003) "Natural Products as Sources of New Drugs over the Period 1981–2002", *Journal of Natural Products* 66: 1022–1037.

Nimmer, Raymond T. (1998) "Breaking Barriers: The Relationship Between Intellectual Property and Contract Law", *Berkeley Journal of Technology Law* 13(245): 245–279.

Office of Technological Assessment (OTA) (1991) *Biotechnology in a Global Economy*, OTA-BA-494, Washington, DC: US Government Printing Office, October, p. 87.

Organisation for Economic Cooperation and Development (OECD) (1995) *Saving Biological Diversity: Economic Incentives*, Paris: OECD.

——— (1999a) *Handbook of Incentive Measures for Diversity: Design and Implementation*, Paris, OECD.

——— (1999b) *Saving Biological Diversity: The Role of Incentives*, Paris: OECD.

Pacón, Ana Maria (1996) "What will TRIPS do for the Developing Countries", in Friedrich-Karl Beier and Gerhard Schricker, eds., *From GATT to TRIPS – The Agreement on Trade Related Aspects of Intellectual Property Rights*, IIC Studies 18, Munich: Max Planck Institute for Foreign and International Patent, Copyright and Competition Law.

Pearce, David and Moran, Katy (1994) *The Economic Value of Biodiversity*, London: IUCN, Earthscan Publications.

Pharmaceutical Research and Manufacturers of America (PhRMA) (2003) *Pharmaceutical Industry Profile 2003*, Washington DC: PhRMA.

Pierce, Alan R. and Laird, Sarah A. (2003) "In search of Comprehensive Standards for Non-Timber Forest Products in the Botanicals Trade", *International Forestry Review*, June, draft version on file with author.

Pindyck, Robert S. (1991) "Irreversibility, Uncertainty and Investment", *Journal of Economic Literature* XXIX(September): 1110–1148.

Plotkin, Mark J. (1991) "Traditional Knowledge of Medicinal Plants: The Search for New Jungle Medicines", in O. Akerele, V. Heywood, and H. Synge, eds., *The Conservation of Medicinal Plants*, Cambridge: Cambridge University Press, pp. 106–116.

Polasky, Steven and Solow, Andrew R. (1995) "On the Value of a Collection of Species", *Journal of Environmental Economics and Management* 29(3): 298–303.

Posey, Darell A. (1996) *Traditional Resource Rights: International Instruments for Protection and Compensation for Indigenous Peoples and Local Communities*, Gland: World Conservation Union (IUCN).

Powell, Walter W. (1996) "Inter-Organizational Collaboration in the Biotechnological Industry", *Journal of Institutional and Theoretical Economics* 152(1): 197–215.

——— (1990) "Neither Market Nor Hierarchy: Network Forms of Organization", in L.L. Cummings and B. Shaw, eds., *Research in Organisational Behaviour* 12 (JAI Press), pp. 295–336.

Powell, Walter W., Koput, Kenneth W. and Smith-Doerr, Laurel (1996) "Inter-Organisational Collaborations and the Focus of Innovation: Networks of Learning in Biotechnology", *Administrative Science Quarterly* 41(1): 116–145.

Principe, Peter P. (1990) "The Economic Significance of Plants and their Constituents as Drugs", in H. Wagner, H. Hikino, and Norman Farnsworth, eds., *Economic and Medicinal Plant Research*, London: Academic Press, pp. 1–17.

——— (1995) "Monetizing the Pharmacological Benefits of Plants", in Michael J. Balick, Elaine Elisabethsky, Sarah A. Laird, eds., *The Medicinal Resources of Tropical Forests*, New York: Columbia University Press, pp. 191–218.

Rai, Arti K. (2001) "The Intersection of Intellectual Property and Antitrust in Cumulative Biopharmaceutical Innovation", presentation given at Conference on Antitrust, Technology and Intellectual Property, Berkeley: Berkeley Centre for Law and Technology, 2–3 March, on file with author.

Raman, Amala and Skett, Paul (1998) "Traditional Remedies and Diabetes Treatment", in H.D.V. Prendergast, N.L. Etkin, D.R. Harris, and J. Houghton, eds., *Plants for Food and Medicine*, Kew, UK: Royal Botanical Gardens, pp. 361–372.

Rausser, Gordon C. and Small, Arthur A (2000) "Valuing Research Leads: Bioprospecting and the Conservation of Genetic Resources", *Journal of Political Economy* 108(1): 173–206.

Reddy, Prasada (2003) "Innovation System of Biopharmaceutical Industry in India", paper presented at the International Workshop on Building (Bio) Pharmaceutical Systems in Developing Countries, United Nations University Institute for New Technologies, Maastricht, The Netherlands, 26–27 February.

Reid, J. (1983) *Sorcerers and Healing Spirits*, Canberra: Australian National University Press, cited in Huft (1995).

Reid, W.V. and Laird Sarah A. (1996) "Biodiversity Prospecting", in Michael J. Balick, Elaine Elisabetsky, and Sarah A. Laird, eds., *Medicinal Resources of the Tropical Forests: Biodiversity and its Importance to Human Health*, New York: Columbia University Press, pp. 142–173.

Reid, W.V., Laird, S.A., and Meyer, C.A. (1996) "Biodiversity Prospecting", in Michael J. Balick, and Elaine Elisabetsky, and Sarah A. Laird, eds., *Medicinal Resources of the Tropical Forest*, New York: Columbia University Press, pp. 42–79.

Reid, Walter V., Laird, Sarah A., Meyer, Carrie A., et al., eds. (1993) *Biodiversity Prospecting: Using Genetic Resources for Sustainable Development*, Washington, DC: World Resources Institute.

Rodney, J.Y. Ho and Garibaldi, Milo (2002) *Biotechnology and Biopharmaceuticals*, New Jersey: Wiley and Sons.

Rogerson, William P. (1992) "Contractual Solutions to the Hold-Up Problem", *Review of Economic Studies* 59(4): 777–794.

Rosenberg, Nathan (1982) *Inside the Black Box: Technology and Economics*, Cambridge: Cambridge University Press.

Rosenthal, J.P. (1997) "Equitable Sharing of Biodiversity Benefits: Agreements on Genetic Resources", in OECD, *Investing in Biodiversity: The Cairns Conference*, Paris: OECD, pp. 253–274.

——— (1998) "The International Cooperative Biodiversity Groups (ICBG) Program", in Case Studies on Benefit-Sharing Arrangements, Conference of the Parties to the Convention on Biological Diversity, 4th Meeting, Bratislava, May.

Rothschild, M. and Stiglitz, Joseph (1976) "Equilibrium in Competitive Insurance Markets: An Essay on the Economics of Imperfect Information", *Quarterly Journal of Economics* 95(4): 629–649.

Sallenave, John (1994) "Giving Traditional Ecological Knowledge its Rightful Place in Environmental Assessment", CARC-Northern Perspectives 22(1), available at http://www.carc.org/pubs/

Saviotti, Pier Paolo (1998) "Industrial Structure and the Dynamics of Knowledge Generation in Biotechnology", in Jacqueline Senker and Ronald van Vliet, eds., *Biotechnology and Competitive Advantage: Europe's Firms and the US Challenge*, Cheltenham: Edward Elgar, pp. 19–44.

Schaaf, Thomas (2000) "Environmental Conservation Based on Sacred Sites", in Darrell A. Posey and Oxford Centre for the Environment, Ethics and Society, eds., *Cultural and Spiritual Values of Biodiversity*, United Nations Environment Programme (UNEP), Intermediate Technology Publications, pp. 325–344.

Scherer, F.M. and Ross, David (1990) *Industrial Market Structure and Economic Performance*, 3rd edn, Boston: Houghton Mifflin, pp. 613–660.

Schuster, B.G., Jackson, J.E., Obijiofor, C.N., et al. (1999) "Drug Development and Conservation of Biodiversity in West and Central Africa: A Model for Collaboration with Indigenous People", *Pharmaceutical Biology* 37(Supplement): 84–99.

Scotchmer, Suzanne (1999) "Cumulative Innovation in Theory and Practice", Working Paper Series, Berkeley: University of Berkeley.

Shiva, Vandana (1997) *Biopiracy: The Plunder of Nature and Knowledge*, Cambridge MA: South End Press.

Simon, Herpert A. (1961) [1945] *Administrative Behavior*, 2nd edition, New York: Macmillan Press.

Simpson, David R. and Sedjo, Roger A. (2004) "Golden Rules of Economies Yet To Strike Prospectors", *Nature* 430(12 August): 723 (www.nature.com/nature).

Simpson, David R., Sedjo, Roger A. and Ried, John W. (1996) "Valuing Biodiversity for Use in Pharmaceutical Research", *Journal of Political Economy* 104(1): 163–185.

Sittenfeld, A., Tamayo, G., Nielsen, V., et al. (1999) "Costa Rican International Cooperative Biodiversity Group: Using Insects and Other Arthopods in Biodiversity Prospecting", *Pharmaceutical Biology* 37(Supplement): 55–68.

Spence, Michael (1973) "Job Market Signaling", *Quarterly Journal of Economics* 87(3): 355–374.

——— (1974) *Market Signalling*, Cambridge, London: Harvard University Press.

Spier, K.E. (1992) "Incomplete Contracts and Signalling", *RAND Journal of Economics* 23(3, Autumn): 432–443.

Stenson, Anthony and Gray, Tim F. (1999) *The Politics of Genetic Resource Control*, Houndmills and New York: Macmillan and St. Martin's Press.

Straus, Joseph (1993) "The Rio Biodiversity Convention and Intellectual Property", *International Review of Industrial Property and Copyright Law* 24(5).

——— (1996) "Implications of the TRIPS Agreement in the Field of Patent Law", in Karl Beier and Gerhard Schricker, eds., *From GATT to TRIPS – The Agreement on Trade Related Aspects of Intellectual Property Rights*, IIC Studies 18, Munich: Max Planck Institute for Foreign and international Patent, Copyright and Competition Law.

——— (1998) "Biodiversity and Intellectual Property", 35th Congress of the Workshop of the International Association for the Protection of Intellectual Property, Workshops I–VII, Rio de Janeiro, International Association for the Protection of Industrial Property, 24–29 May.

Svarstad, Hanne (2000) "Reciprocity, Biopiracy, Heroes, Villains and Victims", in Hanne Svarstad and Shivcharn S. Dhillion, eds., *Responding to Bioprospect-*

ing: From Biodiversity in the South to Medicines in the North, Oslo: Spartacus Vorlag AS, pp. 19–36.

Swanson, Timothy (1995) "The appropriation of evaluation's values an institutional analysis of intellectual property regimes and biodiversity conservation", in Timothy Swanson, *Intellectual Property Rights and Biodiversity Conservation: an Interdisciplinary Analysis of the Values of Medicinal Plants*, Cambridge: Cambridge University Press, pp. 139–141.

Symes, David and Crean, Kevin (1995) "Privatization on the Commons: The Introduction of Individual Transferable Quotas in Developed Fisheries", *Geoforum* 26(2): 175–85.

Tansey, Geoff (1992) "Trade, Intellectual Property, Food and Biodiversity", Discussion Paper, London: Quaker United Nations Office (QUNO).

——— (2002) "Food Security, Biotechnology and Intellectual Property, Unpacking some Issues around TRIPS", Discussion Paper, Geneva: Quaker United Nations Office (QUNO).

Taylor, N. (1965) *Plant Drugs That Changed The World*, London: George Allen & Unwin.

Teece, David J. (1987) "Profiting From Technological Innovation", in David J. Teece, ed., *The Competitive Challenge: Strategies for Industrial Innovation and Renewal*, Massachusetts: Cambridge University Press, pp. 185–219.

Ten Kate, Kerry and Laird, Sarah A. (1999) *The Commercial Use of Biodiversity: Access to Genetic Resources and Benefit-Sharing*, London: Earthscan Publications.

Ten Kate, Kerry and Wells, Adrian (1998) "The Access and Benefit-Sharing Policies of the United States National Cancer Institute: a Comparative Account of the Discovery and Development of the Drugs Calanolide and Topotecan", Benefit-Sharing Case Study, Submission to the Executive Secretary of the Convention on Biological Diversity by the Royal Botanic Gardens, Kew, UK.

——— (2001) "Preparing a National Strategy on Access to Genetic Resources and Benefit Sharing", a pilot study, Kew: Royal Botanic Gardens.

Timmerman, Barbara N. (1997) "Biodiversity Prospecting and Models for Collections of Resources: the NIH/NSF/USAID Model", in K.E. Hoagland and A.Y. Rossman, eds., *Global Genetic Resources: Access, Ownership and Intellectual Property Rights*, Washington DC: Association of Systematics Collections, pp. 219–302.

——— (1999) "The Latin American ICBG: The First Five Years", *Pharmaceutical Biology* 37(Supplement): 35–54.

Tirole, J. (1988) *The Theory of Industrial Organisation*, Cambridge, MA: MIT Press.

Tobin, Brendan (2002) "Biodiversity Prospecting Contracts: the Search for Equitable Agreements", in Sarah A. Laird, ed., *Biodiversity and Traditional Knowledge: Equitable Partnerships in Practice*, London/Sterling VA: Earthscan Publications, pp. 287–309.

Tobin, Brendan and Johnston, Sam (2004) "The International Regime on Access and Benefit Sharing for Genetic Resources and Associated Traditional Knowledge", draft, on file with the author.

UNCTAD (2000) "Report of the Expert Meeting on Systems and National Experiences for the Protection of Traditional Knowledge, Innovations and Practices", Geneva, Trade and Development Board, Commission on Trade in Goods and Service, and Commodities, Fifth Session, 6 December.

——— (2001) "The Sustainable Use of Biological Resources: Systems and National Experiences for the Protection of Traditional Knowledge, Innovations and Practices", Geneva, Trade and Development Board, Commission on Trade in Goods and Service, and Commodities, Fifth Session, 19–23 February.

UNU/IAS (2004) "The Role of Registers and Databases in the Protection of Traditional Knowledge: A Comparative Analysis", UNU/IAS Report, Japan, January.

Verma, Surinder (1998) "Protection of Trade Secrets Under the TRIPS Agreement, and Developing Countries", *The Journal of World Intellectual Property* 1(5): 723–742.

Vlietinck, J. and van den Berghe, D.A. (1998) "Leads for Antivirals from Traditional Medicines", in H.D.V. Prendergast, N.L. Etkin, D.R. Harris, and J. Houghton, eds., *Plants for Food and Medicine*, Kew, UK: Royal Botanical Gardens.

Vogel, Joseph Henry (1996) "Bioprospecting and the Justification for a Cartel", *Bulletin of the Working on Traditional Resource Rights* (4): 16–17.

——— (1997) "Bioprospecting," Biopolicy Journal 2(Paper 5): PY97005.

Walsh, Gary (2003) *Biopharmaceuticals: Biochemistry and Biotechnology*, Chichester UK: Wiley.

Warren, Michael (1991) "Using Indigenous Knowledge in Agricultural Development", World Bank Discussion Paper no. 127, Washington DC: World Bank.

——— (1998) *Indigenous Knowledge and Development Monitor* 6(3): 13.

Watal, Jayashree (2001) "Workshop on Differential Pricing and Financing of Essential Drugs", background note prepared by J. Watal, Consultant to WTO Secretariat, Geneva: World Trade Organization.

Williamson, Oliver E. (1975) *Markets and Hierarchies: Analysis and Antitrust Implications*, New York, Free Press.

——— (1979) "Transaction-Cost Economics: The Governance of Contractual Relations", *Journal of Law and Economics* 22(2): 233–261.

——— (1983) "Credible Commitments: Using Hostages to Support Exchange", *The American Economic Review* 73(4, September): 519–540.

——— (1985) *The Economic Institutions of Capitalism*, New York: Free Press.

——— (1991) "Comparative Economic Organization: The Analysis of Discrete Structural Alternatives", *Administrative Science Quarterly* 36: 269–296.

——— (2000) "The New Institutional Economics: Taking Stock, Looking Ahead", *Journal of Economic Literature* 38(3, September): 595–613.

——— (2001) "The Science of Contract: Private Ordering", August (draft version, on file with author).

Winter, Sidney G. (1987) "Knowledge and Competence as Strategic Assets", in David J. Teece, ed., *The Competitive Challenge: Strategies for Industrial Innovation and Renewal*, Cambridge: Cambridge University Press, pp. 159–183.

——— (1989) "Patents in Complex Contents: Incentives and Effectiveness", in

Vivian Weil and John W. Snapper, eds., *Owning Scientific and Technical Information: Values and Ethical Information*, New Brunswick: Rutgers University Press, pp. 41–60.

Wolfrum, Rüdiger, Klepper, Gernot, Stoll, Peter Tobias, and Franck, Stephanie (2001) "Implementing the Convention on Biological Diversity: Analysis of the Links to Intellectual Property and the International System for the Protection of Intellectual Property", UNEP/CBD/WG-ABS/1/INF/3, available at: www.biodiv.org/doc/meetings/abs/abswg-01/information/abswg-01-inf-03-en.pdf.

Wood Sheldon, Jennie and Balick, Michael J. (1995) "Ethnobotany and the Search for Balance Between Use and Conservation", in Timothy Swanson, ed., *Intellectual Property Rights and Biodiversity Conservation: An Interdisciplinary Analysis of the Values of Medicinal Plants*, Cambridge: Cambridge University Press, pp. 45–64.

Wood Sheldon, Jennie, Balick, Michael J. and Laird, Sarah A. (1997) "Medicinal Plants: Can Utilization and Conservation Co-Exist?", *Advances in Economic Botany* 12, Bronx, NY: New York Botanical Garden.

World Intellectual Property Organization (WIPO) (1998–1999) *Intellectual Property Needs and Expectations of Traditional Knowledge Holders*, Geneva: WIPO Report on Fact-Finding Missions on Intellectual Property and Traditional Knowledge.

——— (WIPO) (2001) *Progress Report on the Status of Traditional Knowledge as Prior Art*, Intergovernmental Committee on Intellectual Property and Genetic Resources, Traditional Knowledge and Folklore, Second Session, WIPO/GRTKF/IC/2/6, Geneva: WIPO.

World Resources Institute (WRI), IUCN, UNEP, et al. (1992) *Global Biodiversity Strategy: Guidelines for Action to Save, Study and Use Earth's Biotic Wealth Sustainably and Suitably*, Washington DC: World Resources Institute (WRI).

Wright, Brian (2000) "Intellectual Property Rights Challenges and International Research Collaborations in Agricultural Biotechnology", in M. Qaim, A.F. Krattinger, and J. von Braun, eds., *Agricultural Biotechnology in Developing Countries: Towards Optimizing the Benefits for the Poor*, Boston: Kluwer, pp. 289–314.

Wu Ning (1998) "Indigenous Knowledge of Yak Breeding and Cross Breeding Among Nomads in Western China", *Indigenous Knowledge and Development Monitor* 6(1): 7.

Zeckhauser, Richard (1996) "The Challenge of Contracting for Technological Information", *Proceedings of the National Academy of Sciences of the United States of America* 93(November): 12743–12748.

Index